T5-BPY-624

ADVANCES IN EDUCATIONAL PRODUCTIVITY

EVALUATION RESEARCH FOR EDUCATIONAL PRODUCTIVITY

Series Editor: HERBERT J. WALBERG
College of Education
University of Illinois at Chicago

Volume Editors: ARTHUR J. REYNOLDS
School of Social Work and
Institute for Research on Poverty
University of Wisconsin-Madison

HERBERT J. WALBERG
College of Education
University of Illinois at Chicago

VOLUME 7 • 1998

 JAI PRESS INC.

Greenwich, Connecticut   *London, England*

001007

Copyright © 1998 JAI PRESS INC.
55 Old Post Road, No. 2
Greenwich, Connecticut 06836

JAI PRESS LTD.
38 Tavistock Street
Covent Garden
London WC2E 7PB
England

All rights reserved. No part of this publication may be reproduced,
stored on a retrieval system, or transmitted in any form or by any
means, electronic, mechanical, photocopying, filming, recording, or
otherwise, without prior permission in writing from the publisher.

ISBN: 0-7623-0253-4

Manufactured in the United States of America

ADVANCES IN EDUCATIONAL
PRODUCTIVITY

Volume 7 • 1998

EVALUATION RESEARCH FOR
EDUCATIONAL PRODUCTIVITY

CONTENTS

0041007

PREFACE

Herbert J. Walberg

This volume continues an annual series of works on educational productivity centered on how more can be accomplished in education without consuming unnecessary human, economic, and social resources. Modern nations seek to enlarge learning as an end in its own right and also to increase "human capital" as a means of increasing individual opportunity for advancement, national economic growth, human welfare, and the quality of life. During the past century, education expanded quantitatively by providing more years of education to ever greater numbers of students. Throughout the world, the fractions of national populations successively completing primary, secondary, and tertiary programs have increased enormously. Today, education is expanding in the form of pre-primary programs on one hand, and, on the other, postsecondary education, on the job training, recurrent education, leisure and retirement courses, and other programs for adults. Today also, we realize the decisive educational influence of the larger world outside the conventional classroom, especially as exercised through the family, the media and the community.

Those concerned with education cannot remain content with mere quantitative expansion. In an age of rapidly expanding knowledge and world economic competition, they must be concerned not only with diplomas and degrees but also how much students have actually learned and the degree to which their knowledge, skills, and attitudes have prepared them for occupations, civic participation, and constructive leisure pursuits. They must be concerned also with the quality of the educational experience and the conservation of human time in learning since our days on earth are limited.

For these reasons, educators must examine curricula, and ask what knowledge is of greatest worth. They must avoid teaching what students already know and what they are yet incapable of learning. They must also ask what teaching methods, and instructional media and systems accomplish the most in the shortest periods of time at least cost. Finally, they must ask about the most efficient ways of governing, financing and organizing education.

To respond to these challenges, educators cannot rely on a single discipline. Anthropology, economics, history, political science, psychology, and sociology— all have useful insights and facts that can be brought to bear. In addition, innovations in practice often outrun what academia is able to systematize; educators need to be informed of efficient breakthroughs wherever they originate.

The volume is the seventh in the annual series "Advances in Educational Productivity". Previous volumes in the series have concerned cost-benefit analysis, organizational effects on educational outcomes, educational evaluation in developing countries, educational cost accounting, and other topics. The theme of "Evaluation Research for Educational Productivity" is that educational and social programs must be evaluated systematically, validly, and regularly to promote educational productivity for students and schools alike.

Chapter 1

INTRODUCTION AND OVERVIEW

Arthur J. Reynolds and Herbert J. Walberg

Evaluation research today is more diverse than ever before. Scanning the pages of journals such as *Evaluation Practice*, *Evaluation Review*, and *New Directions for Evaluation* reveals that researchers are employing a widening array of techniques and approaches to advance knowledge about the effects of educational and social programs, and to inform policy discussions at the local, state, and national levels. The demand for knowledge about programs is growing even if the resources to produce it are not. Evaluation researchers and stakeholders are asking not only if programs are effective but if particular programs are more effective and efficient than others? They want to know how and why programs work, and for whom do they work best. How well, they ask, are programs implemented? How well are they conceptualized? How generalizable are findings from one study or a group of studies? Do different stakeholders view the effectiveness of programs similarly?

   The purpose of this volume is to bring together work by evaluation researchers who employ a broad variety of contemporary methods to answer these and other questions. Rossi and Freeman (1993) defined evaluation research as the "systematic application of social research procedures in assessing the conceptualization and design, implementation, and utility of social intervention programs" (p. 5). Although our coverage is not exhaustive of all current methods, most emerging

Advances in Educational Productivity, Volume 7, pages 1-12.
Copyright © 1998 by JAI Press Inc.
All rights of reproduction in any form reserved.
ISBN: 0-7623-0253-4

and prominant evaluation perspectives and methods are included such as theory-driven evaluation, experimental and quasi-experimental approaches, stakeholder-oriented evaluation, process evaluation, qualitative and case-study methods, benefit-cost and cost-effectiveness analysis, and research synthesis. Many of the contributors have already written extensively on these topics, and we envisioned that their chapters would summarize the goals of a particular evaluation method and how it can be best applied in actual evaluations.

As suggested above, the volume addresses evaluation research broadly and includes traditional evaluations of programs developed by psychologists and educators for local settings as well as analyses of government-initiated educational policies and general social conditions with origins in economics and sociology. Consequently, the chapters include a wide range of program evaluations and policy analyses. The volume also is eclectic, exemplifying the history, current status, and future of evaluation research. Moreover, the coverage of topics in the chapters is multidisciplinary spanning the disciplines of economics, education, human development, psychology, sociology, social welfare, and statistics.

We wanted to assemble a set of exemplary chapters usable by people who are not specialists in the topics of each chapter but who are interested in employing the ideas and methods in evaluations and policy analyses. We hoped that, as consumers of evaluation research, program administrators and policy-makers would profit from the chapters, and use, if not the techniques themselves, many of the concepts and perspectives. For these reasons, we encouraged reflective accounts about the methodologies and lessons learned from what went right and wrong in actual efforts. A key question we wanted to see addressed is: How do current developments in evaluation research enhance our capacity to come to conclusions useful to policymakers and program professionals? Many of the chapters address this question, and Carol Weiss's chapter is entirely devoted to it. Finally, the volume may be particularly beneficial to students of evaluation research as well as practitioners in keeping abreast of current trends in this multidisciplinary profession that has been described as the "worldly science" (Cook & Shadish, 1986).

This volume may help evaluators address two recurrent challenges that may be best viewed as opportunities: (1) evaluation use and (2) investment. First, it is our view that the increasing diversity of evaluation has not resulted in significantly greater use of evaluation findings by program administrators and policymakers. This is partly attributable to the fact that, along with being diverse, the evaluation profession is fractured. Evaluators often spend a significant amount of time and effort justifying particular evaluation approaches or defending specific evaluation findings, and proportionately less time on dissemination and utilization of findings. In recent years, interest has increased in encouraging integrative evaluation practices among quantitatively and qualitatively-oriented evaluators (e.g., Reichardt & Rallis, 1994; Shadish, Newman, Scheirer, & Wye, 1995). In this volume, the chapters provide many illustrations of how "state of the art" evaluation practices can address important educational and social issues, and promote knowledge transfer, program improvement, and implementation of integrative evaluation practices.

A second challenge this volume can help address is the area of evaluation investment. Although evaluation training has improved over the last two decades, especially with the advent of meta-analysis, the availability of sophisticated statistical techniques, and the wider use of qualitative methods, financial investment in evaluation research on educational programs is relatively low, at least compared to historical patterns. The Head Start Preschool Program, for example, expended only 0.3 percent of its 1996 federal expenditure on research and evaluation whereas 2.5 percent of it's expenditure in 1974 was spent on research and evaluation (U. S. General Accounting Office, 1997). Given the high funding priority for this program, this level of investment in evaluation may be higher than for most other programs. The increasing devolution of programs through block grants to the states provides many opportunities for evaluators to contribute to the planning and analysis of new programs and to demonstrate to administrators and policymakers the value of evaluation information.

Through describing the vibrancy of contemporary evaluation approaches and their relation to program improvement activities, investments in evaluation may be better justified. In this volume, we try to go beyond debates about appropriate philosophies, competing approaches, and "insider" discussions to show why investment in evaluation research is valuable and how evaluation can contribute to educational productivity. The chapters provide many examples of evaluations that have done just this. Indeed, current developments in such areas as utilization-focused evaluation (Patton, 1997), theory-driven evaluation (Chen, 1990), and meta-analysis (Cooper & Hedges, 1994) suggest that evaluation can contribute more to policy analysis and program improvement in the future.

OVERVIEW

In the rest of this chapter, we describe the organization of the volume and the contributions of the individual chapters. The volume is organized in three sections. The first is Evaluation Planning and Design. A great deal of attention should be devoted to the design of a program evaluation, much more than usually occurs. Understanding the theory of the program, identifying key stakeholders, and planning the research design to strengthen the causal interpretation of findings are critical features. The chapters discuss the purposes and utility of theory-driven evaluation, quasi-experimental designs, cross-design synthesis, and stakeholder-based perspectives.

The second section is Implementation Process and Contexts. Knowledge about the delivery of program services and the context in which a program is implemented is prerequisite to impact assessment. The quality with which programs are implemented and the perspectives of participants about their program experiences and the values they hold can provide evaluation researchers with a more complete understanding of program efficacy. The three chapters in this section consider contemporary approaches to conducting implementation evaluation, qualitative evaluation, and case studies.

The final section is Analysis and Utilization. A growing number of analytic techniques are available to the evaluation researcher in helping to analyze data and to conduct impact assessments. Although many of these techniques require advanced statistical training or collaboration, they are important for strengthening confidence and utilization of findings. In five chapters, the authors discuss "state of the art" developments in statistical methods, research synthesis (meta-analysis), cost-benefit analysis and cost-effectiveness analysis, secondary analysis, and evaluation uses for policy and program development.

The organization of this book highlights typical stages of activities in conducting evaluations. Indeed, a particular evaluation approach may be best understood within the context of the stage in which it is practiced. Some evaluation techniques described, however, overlap these sections. Theory-driven evaluation, for example, can be used to conduct an implementation or impact assessment just as it can to inform evaluation planning efforts. Case study and ethnographic methods are not used only for conducting process and implementation evaluations but also have much to contribute to data analysis and utilization. These approaches, however, are particularly valuable for providing "grounded" knowledge about program operations and understanding. In sum, the methods, practices and perspectives described provide a comprehensive set of tools for addressing many evaluation questions. While almost all evaluation approaches are discussed to some extent, detailed coverage of many topics (e.g., needs assessment, advanced statistical methods such as structural modeling and hierarchical linear models) was necessarily limited. Readers may wish to consult the references to complementary works and examples cited at the end of the chapters.

PART I. EVALUATION PLANNING AND DESIGN

Theory-Driven Evaluations

This section begins with "Theory-Driven Evaluations" by Huey-tsyh Chen. Evaluation planning should start with a full understanding of the program theory and how the organization and delivery of the program should serve the intended target population. Program theory is defined as "...a specification of what must be done to achieve the desired goals, and other important impacts may also be anticipated, and how these goals and impacts would be generated." As Huey-tsyh Chen points out, educational programs, like other action programs function under a set of assumptions regarding how the program should operate and what mechanisms are required for the program to be effective. These assumptions and mechanisms are the major focus of theory-driven evaluations since they are employed by program designers and implementors.

Chen's chapter shows how theory-driven evaluations provide useful information not only to assess the merit of the program, but also to improve the program. His chapter introduces the concepts of theory-driven evaluations and discusses why they are useful in assessing educational productivity. Chen explains the different types of theory-driven evaluations and how to carry them out. He also identifies three dimensions that are important for selecting a particular research method to implement theory-driven evaluation: depth versus width of information needed in an evaluation context, high versus low availability of data, and high versus low openness of environmental influences (i.e., level of control available to the researcher). Chen also provides examples of normative, causative, and mixed-type evaluations.

Design Elements of Quasi-Experiments

In "Design Elements of Quasi-Experiments," William J. Corrin and Thomas D. Cook discuss the advantages of careful planning of evaluation designs to reduce causal uncertainty and facilitate the interpretation of findings. The authors address issues related to the use of experiments and quasi-experiments for identifying the causal impacts of education programs. Experiments, of course, require random assignment of students, classrooms, or schools to different programs to ascertain their respective merits or to see how any one of them fares compared to no program. When experiments cannot be carried out or when it seems advisable to avoid the special conditions they require, evaluators employ quasi-experiments. Quasi-experiments are studies in which participation in alternative programs is not randomly assigned and thus program assignment may well be confounded with differences between the populations or samples being contrasted. In this chapter, the authors advocate thinking out quasi-experiments in terms of design elements rather than in the way they have been most often presented in the past—as a circumscribed set of individual quasi-experimental designs.

As Corrin and Cook point out, quasi-experimental designs are typically chosen on the basis of their feasibility and the degree of certainty stakeholders require about whether the program actually causes observed changes. Good designs benefit evaluators because they improve the certainty with which they can make causal claims. These designs also reduce dependence on statistical adjustment during subsequent data analysis. By focusing on design elements rather than on pre-defined designs in planning their research, evaluators gain the flexibility to adapt pre-defined designs or create new designs to their specific research situations while maintaining the benefits of quasi-experiments in promoting causal inference. Thus, the thoughtful combination of design elements enables evaluators to address issues of causal description and match their designs with their particular research contexts.

Cross-Design Synthesis

Closely related to quasi-experiments is the subject of Robert F. Boruch and George Terhanian's chapter "Cross-Design Synthesis." They focus on improving education production function research through cross-design synthesis. This new approach combines estimates of effects of experiments and of national sample surveys, each having a unique strength. By integrating different methodological approaches, the validity and generalizablity of estimates of program effects can be strengthened. Their views are particularly pertinent to federal institutional policy for generating data for productivity research. They lay out rationale for our approach based on scientific political-institutional standards.

Boruch and Terhanian give two examples of cross-design synthesis for better understanding the effects of adult literacy programs in the United States and for strengthening the validity of findings concerning the effects of ability grouping on school achievement. Both examples illustrate how cross-design synthesis can improve the analysis of research on production functions, and thus enhance the implications of evaluation findings for policy development.

Stakeholder Concepts

Who pays attention to evaluations? Do they have any effect on policymakers, program designers, and clients? In their chapter, Marvin Alkin, Carolyn Hofstetter, and Xiaoxia Ai discuss evaluation from the perspective of different stakeholders and how such perspectives influence the design, conduct, and impact of evaluations. They hold that all those that have a "stake" in evaluations should be involved but find that implementing this ideal is challenging. Even so, successful implementation improves the evaluation's relevance the several parties, stakeholders' commitments to the evaluation, and the applicability of the evaluation.

Alkin and colleagues specify the number and roles of stakeholders and explains the basis for choosing them. They point out that in any given evaluation, stakeholders may help identify the problem, participate in the design and methodology of the evaluation, furnish and interpret data, and aid in the design and completion of the evaluation report. In these ways, Alkin and colleagues find that stakeholder evaluation incorporates such desirable features as responsiveness to perceived problems, focus on utilization, and participatory involvement.

PART II. IMPLEMENTATION PROCESS AND CONTEXTS

Evaluating Implementation

Once planning is completed, an essential part of any evaluation is to document and verify program implementation. Implementation or process evaluations are underutilized, which is ironic given that many evaluation theorists believe that the modest record of social programs in meeting their objectives is attributable to poor program implementation. Ann R. McCoy and Arthur J. Reynolds show how process evaluation can document the chain of events in implementation of a program. The documentation shows what the components of the intervention were, how they were delivered, and who received them. Information obtained through implementation evaluation facilitates program dissemination thereby encouraging its future use by others.

Among the questions the authors address are (i) what is implementation or process evaluation, (ii) when is process evaluation necessary, and (iii) how is process measured for evaluation purposes? They illustrate the importance of process evaluation through research on the well known Head Start preschool program and James Comer's School Development Program. The first case shows the dangers of not conducting evaluation of program implementation prior to outcome evaluations. Indeed, the first national evaluation of Head Start did not verify that the program services were delivered as intended and with sufficient quality. The second case illustrates the value of conducting ongoing evaluations of implementation for several years prior to formal outcome evaluation. Program staff needed several years to debug the program and implement it smoothly to maximize the value of the later outcome evaluations.

Qualitative, Interpretive Evaluation

Qualitative evaluations of educational programs offer a significant counterpoint to large-scale comparisons of standardized performance of learners inside and outside a given program. As Jennifer C. Greene points out, the drama of narratives enables policymakers and program developers to understand workings of programs for particular learners in particular sites and to understand individual experiences, as her three detailed examples show. Each example (i.e., teacher development and school reform, the Head Start preschool experience, and the meaning of "at risk") illustrates the fundamental importance of the social context in which programs are implemented and the meanings that participants attach to them.

Qualitative evaluators have challenged policy and program assumptions that poorly connect to individual experiences; they have insisted that unstated values be examined critically and pluralistically, as further examples illustrate. The sensibility of an educational policy "on the ground" remains a unique contribution of qualitative evaluation to advancing educational productivity. This grounded perspective can even re-frame the very concept of productivity. As Greene concludes, the challenge facing qualitative evaluation is to speak with more authority and to be heard with greater credibility.

Case Study Methods

Case study methods offer another valuable perspective for understanding the contexts in which programs are implemented, the program implementation process, and the experiential knowledge about program quality and practices, all of which are important for valid inferences about program impact and generalizability. As Linda Mabry discusses in her chapter, every program evaluation can be considered a case study and it is the responsibility of the evaluator to gain knowledge of the program and its workings in order to optimize program understanding, and ultimately, to promote program improvement. As indicated, case study methodology can add "distinctive depth through presentation of the multiple perspectives of insiders and through opportunities for personal, experiential understanding."

Mabry provides several examples of how case studies promote a deeper understanding of program operations that are difficult to obtain with exclusive reliance on other methods. Indeed, the social embeddedness of educational and social programs, like that of human behavior, seems to require greater utilization of case studies in evaluation research. As discussed, case study methods can be incorporated into an evaluation in many ways spanning both intensive investigation of a particular program site or as part of a multi-method design of a large-scale evaluation.

PART III. ANALYSIS AND UTILIZATION

Statistical Methods for Real-World Evaluation

Evaluation data often contain serious problems that are seldom covered in statistical texts. David Rindskopf's chapter on "Statistical Methods for Real World Evaluation" covers new statistical methods for dealing with such problems as handling large amounts of data, complex sample designs, missing data and sample attrition, nonrandom assignment, design of measurement instruments, and the presentation of results to diverse audiences. Rindskopf explains the reasons for these methods and provides examples of how they solve problems in analyzing typical data collected in evaluations.

The methods discussed include sensitivity analysis to provide bounds on estimates of treatment effects. To strengthen causal inferences in observational field studies, for example, it is necessary for researchers to examine hidden biases in their data through analysis of the consistency of findings across sites and program outcomes, as well as under different model specifications. Other examples relate to new developments in selection modeling and propensity score analysis (also see the chapter by Borsch and Terhanian in this regard), advantages and disadvantages of different methods of covariance adjustment with measurement error, latent variable causal models, longitudinal data analysis, hierarchical models, meta-analysis, and ways to present evaluation data for optimal effectiveness. A set of references are provided for further reading in several disciplines including psychology, statistics, economics, and sociology.

Research Synthesis

In the study of lung cancer, randomized experiments showed the carcinogenic effects of cigarette smoke on infra-humans, but surveys of people were also helpful in drawing conclusions. In varying circumstances, they showed that even with controls for age, sex, socioeconomic status, and other factors, people who smoke longer and more intensively are more likely to experience lung cancer. In education and social policy, we increasingly face many multivariate evaluations or policy studies, each analyzing sets of similar, possibly causal but non-randomized variables. How can we make quantitative sense of the set of multivariate studies as a whole beyond repeating or verbally summarizing the investigators' conclusions?

To answer such questions, Betsy Jane Becker looks at some diverse research on educational productivity that has arisen in a variety of fields including economics, psychology, and sociology. Further increasing complications, educational productivity has also been measured in many ways and at different levels of the educational system including the individual student, classroom, and school. Similarly, different studies have examined a variety of predictors of educational outcomes.

Research synthesists attempt to make sense of such disparate, hard won data and to detect causal relations that pervade the varying data sets and analyses. As an example, Becker reviews past syntheses of studies of multivariate educational productivity. She then describes and illustrates new methods of synthesis that extend meta-analyses of experiments and quasi-experiments to cases of multiple independent variables. She points out that such methods could be widely employed to synthesize many evaluations and policy studies in a variety of areas.

Evaluation Using Secondary Data

Just as Boruch and Terhanian call for better designed survey data, Becker makes the case for more synthesis of existing studies. Similarly, Judy A. Temple recommends better analysis of existing data. Research using administrative and survey data is a cost-effective means of evaluating educational conditions, programs, and policies. The availability of large-scale and often longitudinal data sets offers opportunities for education researchers to obtain policy-relevant research findings without going through the expensive and time-consuming process of original data collection.

Temple's chapter explains how secondary data analysis can enhance researchers' capacities to come to conclusions useful to policy makers and program professionals. Temple discusses the availability of data from the National Center on Education Statistics, the National Science Foundation, and other sources (including other researchers). She describes a number of important issues that arise in estimation and interpretation of results. Temple discusses recent databases from a number of social science disciplines to show how researchers have used secondary data to evaluate educational programs and policies including compensatory preschool education, ability grouping, and private versus public schools. Greater use of existing data to address socially important topics may promote greater investment in evaluation.

Benefit-Cost Analysis and Related Techniques

Benefit-cost analysis is of keen interest to policymakers but is implemented far less often than other methods. In "Benefit-Cost Analysis and Related Techniques," W. Steven Barnett provides an introduction to benefit-cost analysis and the associated techniques of cost analysis and cost-effectiveness analysis. Cost analysis is a full accounting of the resources needed to implement a program. Cost-effectiveness analysis is used to compare alternative approaches to achieving the same outcome objective (e.g., higher reading achievement). In benefit-cost analysis, dollar values are used to index effect sizes across many program outcomes. As Barnett notes in the beginning of the chapter, the logic of benefit-cost analysis is straightforward and nearly all individuals make decisions every day through their own appraisal of costs and benefits. In clear and lucid language, Dr. Barnett discusses how these economic analyses are conducted, their advantages and limitations, and how they can contribute to program evaluation, educational decision making, and ultimately educational productivity.

To illustrate the contributions of these techniques, several practical examples are provided. A cost-effectiveness analysis of four primary school reforms indicated that the annual cost of producing a one-month gain in reading achievement was lowest for peer tutoring relative to adult tutoring, computer-assisted instruction, and extending the school day. A more detailed example of benefit-cost analysis was provided using the well-known High/Scope Perry Preschool Program. Analyses through age 27 suggested that program benefits to program participants, taxpayers, and society (mainly through increases in educational attainment and employment and reductions in crime) outweighed program costs. Clearly, increasing the amount and quality of benefit-cost analyses can go a long way in improving policy decisions as well as the utilization of evaluation findings for program improvement.

Improving the Use of Evaluations

Educational researchers and evaluators expect that the research they produce will help policymakers and administrators make wiser decisions. As Carol H. Weiss finds, they are frequently nonplused that nothing seems to change after their contributions to the debate. If pressed, they may acknowledge that their research does not present the final authoritative word and that evaluation findings at variance with theirs have also been reported. But, as Weiss points out, many other barriers prevent the adoption of research findings including the costs of change, lack of support from policy actors and practitioners, and shortage of personnel with necessary qualifications, which the evaluation has not addressed.

Weiss points out some strategies that evaluators and researchers can use to increase the likelihood that findings will be heeded. They include involving administrators and policymakers in the evaluation process, giving them interim feedback of results so that findings do not come as a surprise at the end, meeting with administrators and staff over time to discuss the importance of findings for the specific conditions that the organization faces and listening carefully to their interpretations and constraints, working with them to develop applications of the findings, undertaking new studies that meet emerging needs, and informing other publics through the general and specialized media about the findings in order to generate communities of support. Weiss also points out that meta-analysis of multiple similar evaluations (see Becker's chapter) and "empowerment" evaluations can also improve the use of evaluations.

CONCLUSION

Though the overviews above bring out important highlights, they hardly substitute for the chapters. For this reason, each chapter in the body of this book deserves careful reading. It can be hoped each will repay the attention of those who seek a beginning understanding of the broad scope of evaluation methods. It can also be hoped that they may serve experienced evaluators who wish to inform themselves of much of the latest thinking in specialties other than their own. Though the authors speak for themselves, some important themes cut across their chapters deserve emphasis. Our concluding chapter identifies these and sets them in the context of the past, present, and the likely future of evaluation research on educational productivity and related topics.

REFERENCES

Cook, T. D., & Shadish, W. R. (1986). Program evaluation: The worldly science. *Annual Review of Psychology, 37*, 193-232.

Chen, H.T. (1990). *Theory-driven evaluations*. Newbury Park, CA: Sage.

Cooper, H. M., & Hedges, L. V. (Eds.). (1994). *The handbook of research synthesis*. New York: Russell Sage Foundation.

Patton, M. Q. (1997). *Utilization-focused evaluation. The new century text, edition 3*. Thousand Oaks, CA: Sage.

Rossi, P. H., & Freeman, H. E. (1993). *Evaluation: A systematic approach* (5th ed.). Newbury Park, CA: Sage.

Reichardt, C. S., & Rallis, S. E. (1994). *The qualitative-quantitative debate: New perspectives*. New Directions for Evaluation (No. 61). San Francisco: Jossey-Bass.

Shadish, W. R., Newman, D. L., Scheirer, M. A., & Wye, C. (1995). *Guiding principles for evaluators*. New Directions for Evaluation (No. 66). San Franciscio: Jossey-Bass.

U. S. General Accounting Office. (1997, April). *Head Start: Research provides little information on impact of current program*. Report HEHS-97-59. Washington, DC: Author.

PART I

EVALUATION PLANNING AND DESIGN

Chapter 2

THEORY-DRIVEN EVALUATIONS

Huey-tsyh Chen

INTRODUCTION:
WHAT IS THEORY-DRIVEN EVALUATIONS

There has been a strong interest in the development of theory-driven evaluations (e.g, Bickman, 1987; Chen, 1990, 1996, 1997; Weiss, 1996, 1997). Much progress has been made, not only in refinements and expansion to the theoretical framework of theory-driven evaluations, but also on the applications of theory-driven evaluations (e.g., Bickman, 1990, 1987; Chen, 1990, 1997a, 1997b; Chen & Rossi, 1992). The purpose of this chapter is to introduce the basic concepts, rationale, and methodology for designing and conducting theory-driven evalua-tions with an emphasis on application so that this evaluation approach could be widely applied in areas such as education.

According to theory-driven evaluations, the foundation to design an evalua-tion is the crucial assumptions underlying an action program regarding how a program is supposedly to be operated and/or why the program is supposed to be effective (Chen, 1990). This kind of evaluation is different from traditional evaluations that rely heavily on research methods as a backbone for designing an evaluation. Before introducing theory-driven evaluations, it is helpful to

Advances in Educational Productivity, Volume 7, pages 15-34.
Copyright © 1998 by JAI Press Inc.
All rights of reproduction in any form reserved.
ISBN: 0-7623-0253-4

discuss traditional method-driven evaluations that evaluators have been familiar with. Under the framework of method-driven evaluations, the design of an
evaluation is mainly guided by the predetermined research steps required by
the application of the particular method (Chen & Rossi, 1992). For example,
when the particular method to be used in a method-driven evaluation is a classic randomized experiment, such as randomization, the major concern in evaluation tends to be focused on activities central to the application of the
experimental designs, such as randomization, before and after measures for
treatment and control groups, the use of analysis of variance for data analysis,
and so on. Similarly, if a naturalistic approach is used, then the evaluation processes tend to focus on activities central to application of naturalistic
approaches, such as observing program operations in the field, making sense of
program implementors' views and ideas, and so on.

There are advantages for doing method-driven evaluations. Since a method-
driven evaluation is guided by a particular method, its evaluation activities are
somewhat standardized. If an evaluator is familiar with research steps of a particular method, then the same research principles can be literally applied to different
types of programs or settings. Method-driven evaluation simplifies evaluation
steps by narrowing down crucial issues central to the application of that particular
method such as ensuring internal validity in the experimental tradition and authen-
ticity in naturalistic approaches. Furthermore, since the method is the central
focus, the quality of the evaluation can be judged in terms of prestige of the particular method used and of how closely the evaluation follows the research principles of that method.

However, it is important to understand its limitations. Method-driven evaluation may have high methodological rigor but may not be comprehensive
enough for providing useful or enlightening information for understanding,
assessing, and/or improving a program. A good example of method-driven evaluation is a black box evaluation in which an outcome evaluation provides information on the relationship between input and output of the program without
providing information on how the input is transformed into output. This kind of
evaluation could use highly sophisticated research designs and statistical models in assessing goal attainment, but provides little information on the following question: What exactly is the treatment or intervention? How the treatment
or intervention is implemented? Whether goals used in evaluation are justifiable or appropriately measured? And why does the treatment or intervention
attain or not attain the goals? Due to lack of contextual information, the results
of a black box evaluation are difficult to interpret or use or could even be misleading (Chen, 1990, 1994b).

An alternative to method-driven evaluations is theory-driven evaluations.
According to the perspective of theory-driven evaluation, evaluators need to have
a conceptual understanding of crucial assumptions underlying a program before
considering or choosing a research method for the evaluation. Since the assump-

tions underlying a program, formally called program theory (Chen, 1990), are a key to understanding theory-driven evaluations, the nature, meanings, and concepts related to these assumptions needs to be discussed in detail. Every action program, including education, is explicitly or implicitly designed and operated under a set of assumptions. Some of these assumptions are prescriptive in nature; others are descriptive in nature. Prescriptive assumptions prescribe what must be done in order to achieve program goals; descriptive assumptions explain why the program can achieve the goals.

Descriptive assumptions are formally called causative theory because these assumptions identify the underlying causal mechanisms that link, mediate, or moderate the treatment or intervention and outcomes. In other words, causative theory explains how the program works by identifying the conditions under which certain processes will arise and what their likely consequence will be (Chen, 1990). More specifically, causative theory covers issues such as the following: What kind of relationships exist between the treatment or intervention and the outcomes? What kinds of factors could be mediating or moderating of the treatment or intervention effect? Under what kind of contextual conditions will the causal relationship be facilitated or inhibited?

For example, in the Performance Contracting Experiment (Gramlich & Kosel, 1975), the program assumes that the intervention could improve students' performance by mechanisms such as that of private firms being offered contracts to teach reading and mathematics and being paid in proportion to the students' gains in test scores. The firms would then produce innovative teaching materials and activities that would enhance students' motivation for learning and their understanding of the subjects which may in turn increase their mathematics and reading scores. Descriptive assumptions emphasize causal relationships or chains between input and output of the program. These assumptions are parallel to the concept of causal theories in social sciences. Weiss (1997) proposes to define program theory mainly based upon the descriptive assumptions.

However, as mentioned earlier, it is important to note that a program also has prescriptive assumptions that prescribe what must be done in order to achieve the predetermined goals. The prescriptive assumptions are formally called normative theory, and this theory prescriptively dictates which goals should be pursued and operationalized, what treatments are needed, and the program environment and organization necessary for implementing treatment and pursuing goals. More specifically, prescriptive assumptions prescribe what goals should be pursued, what treatment or intervention should be, and what implementations are needed in order to achieve the goals. For instance, in the example of Performance Contract Experiment, the prescriptive assumptions prescribe that private firms could develop and provide the innovative instructional materials teaching machines in a timely fashion, could find and hire competent and committed teachers to teach students, could gain cooperation from the schools and regular teachers, and so forth.

Based upon the above discussion, program theory is formally defined by Chen (1990, p. 43) as: "...a specification of what must be done to achieve the desired goals, and other important impacts may also be anticipated, and how these goals and impacts would be generated." This definition consists of both prescriptive and descriptive assumptions. It is highly important for program theory to cover both kinds of assumptions for the following reasons: (a) Prescriptive assumptions represent the energy or the action aspect of a program. Without including prescriptive assumptions, program theory would exclusively focus on causal issues which do not have direct bearing on stakeholders' day-to-day activities. In order for program theory to catch on where the action is, it is important for program theory to include normative theory. (b) The inclusion of both prescriptive and descriptive assumption in program theory allows evaluators to systematically examine relationships between these two set of assumptions and provides rich information for understanding or for improving the program. (c) To confine program theory as to descriptive assumptions tends to make theory-driven evaluation on a format of causal modeling such as path analysis and structural equations models. On the other hand, by integrating normative and causative theories, the framework of the theory-driven evaluation has the capacity for theory-driven evaluation to benefit from different types of research methods: quantitative, qualitative, or mixed (Chen, 1997b).

There are different labels of this kind of theory-oriented evaluations depending on which aspects of program theory are stressed. For example, Rodriguez and June (1997) use the term "theory-oriented evaluation" by stressing the interpretivist aspect of constructing program theory, while Weiss (1977) uses the terms "theory-based evaluation" for denoting the causative aspect of program theory. The term, logical model, or evaluability assessment, is used in the context of emphasizing the sequential steps in the normative theory that are necessary to achieve goals (Wholey, 1994). Regardless of the variation, the principles, strategies, and techniques discussed in this chapter are essential to these theory-oriented evaluations.

CONCEPTUAL FRAMEWORK OF PROGRAM THEORY AND THEORY-DRIVEN EVALUATIONS

Based upon Chen (1990), a conceptual framework of program theory for guiding theory- driven evaluations is illustrated as in Figure 2.1.

The conceptual framework consists of the following major domains and dimensions and the relationships among them:

(1) Treatment or Intervention

Treatment or intervention is the major change agent in a program. It is believed by program designers and other program stakeholders that the implementation of treatment or intervention can lead to alleviate a social problem or to attain some goals. For example, the treatment of a drug abuse treatment pro-

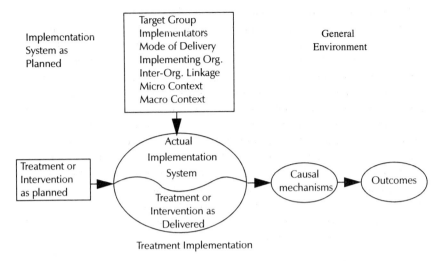

Figure 2.1.   Conceptual Framework of Program Theory

gram may consist of the elements such as individual counseling, Alcoholics Anonymous meetings, and family therapy. Similarly, a state may initiate an anti-drunk driving program by imposing a tough law to punish drunk driving as an intervention to crack down on drunk driving.

(2) Implementation System

   In action programs, another input that is as important as the treatment or intervention is the implementation system which provides means and a context in implementing the treatment or intervention. To implement the treatment or implement the intervention, a program needs to also create an implementing system for doing the following things: attracting target groups, the hiring and training of the program staff, the devising of the mode of delivery, the creation of an organization to coordinate the efforts, and so on. It is vital to recognize that the impacts or consequences of a program is a joint effect of the treatment and implementation system. For example, the success of a job training program is not totally determined by the curriculum, but also by the quality of the teacher, the motivation and attitude of the trainees, the job searching strategies, the local economy, and so on. Similarly, an HIV prevention program for reducing needle sharing among hard-core drug users may consist of a normative theory that assumes that the hired outreach worker can effectively identify and recruit the hard-core drug abusers to attend group therapy sessions. The program contains enough incentives for keeping hard-core abusers participating in most of the sessions. The counselors who conduct these sessions are qualified and enthusiastic.

The implementation system consists of the following dimensions, such as, target group, implementer, mode of delivery, implementing organization, micro context, and macro context (Chen, 1990). These dimensions are concerned with issues such as these: Who is the target group? Is the treatment reaching the target group? Do the implementors possess the required qualities? Are the modes of delivery and organizational coordination appropriate? And so on.

(3) Treatment or Intervention Implementation:

Treatment or intervention implementation relates to the actual delivery of the treatment or intervention by the implements to the target group under the implementation system and the general environment, such as an organization, community, or society. For example, treatment implementation of a new curriculum in a school may mean the process of how the teacher teaches students the new curriculum under the school instructive rules and administrative guidelines. The treatment or intervention as planned may not be exactly the same as the treatment or intervention as the original plan. For example, the teacher may only cover a part of the new text book as original plan or change original curriculum along the way. When the implementation system fails to implement the treatment or intervention as it supposed to be done, it is called an implementation failure.

(4) Causal mechanisms

Theoretically speaking, treatment implementation is a necessary, but not a sufficient condition for program effectiveness. In other words, it is possible that the treatment or intervention is implemented successfully, but the program still fails to achieve its goals. For example, a workfare program may be able to train the welfare recipients the appropriate job skills, but the trainees may still have a difficult time finding a job after graduation. There is mediating moderating factors, such as market demand and corporations' willingness to hire workfare trainees that mediate between job skills and employment. Causal mechanisms refer to those mediating or moderating factors that intervene between the treatment implementation and the anticipated program outcomes. The linkage between the treatment implementation and causal mechanism is called action theory. The linkage between causal mechanisms and goals is called conceptual theory. This conceptualization implies that program failure can result from different sources: implementation, action theory, and conceptual theory. For example, if an HIV prevention outreach program fails to reach the target group, it is an implementation failure. When the target group is reached, but the program fails to enhance the target group's knowledge on risk prevention, it is an action theory failure. Conceptual theory failure happens when the target group is knowledgeable on risk reduction, but the prevention knowledge does not effectively change the target group's high risk sexual behavior.

(5) Outcomes

Outcomes are the result of the transformation. One of the crucial outcomes is the attainment of program goals. The achievement of program goals provides a justification of the existence of a program. For example, the output of an abuser treatment program is whether the abusers stop or continue to abuse their spouse. In addition, a program may also have unindexed outcomes. Unindexed outcomes could be desirable or undesirable. An example of desirable unindexed outcome is that a drug prevention program for youth may bring youths from different ethnic backgrounds for various activities that may reduce the prejudice among groups. An example of undesirable unintended outcomes is that an incarceration of juvenile offenders may provide them opportunity to learn criminal techniques from each other.

(6) General Environment

General environment is the general context such as an organization, community, or society under which the program is getting resources An action program is an open system; that is, it needs to interchange with its environment in order to survive. An action program depends on the environment, such as a community, for its input (money, clients, personnel, social support, etc.) Furthermore, the output of an action program also needs to recognize its environment as valuable and acceptable in order for the program to continue. For example, if the output of a hospital is many dead patients, the hospital will be boycotted by the community or shut down by the government.

The arrows in the diagram of Figure 2.1 suggest that in order for a program to be effective, both the normative and causative theories have to work. In general, for a program to be effective, the treatment or intervention and the implementation system has to be not only sound as specified in the normative theory, but also adequately implemented in the field. Furthermore, the joint force of the treatment or intervention and the implementation system have to successfully activate the causal mechanisms which then affects the outcomes as specified in the causative theory. The framework provides a guidance regarding which domains or relationships among domains needs to be focused on an evaluation. Depending on stakeholders needs and available evaluation resources, a theory-driven evaluation can focus on some or whole aspects of the conceptual framework of program theory.

STRATEGIES FOR FORMULATING PROGRAM THEORY

The design of theory-driven evaluations starts from the construction of program theory.[1] The following are strategies for constructing program theory:

(1) Determine What Aspects of Program Theory are Needed to be Constructed

In formulating program theory, evaluators need to understand stakeholders' evaluation needs in order to determine what aspects of program theory is appropriate for the particular evaluation situation. If the stakeholders' interests and concerns are on issues such as whether the program is implemented as it is supposed to be, the focus will be on constructing a normative theory. If their interests and concerns are on whether or not the underlying causal mechanisms are working, then a causative theory needs to be constructed. If their interests and concerns relate to both prescriptive and descriptive assumptions of the program, then the comprehensive overall program theory that covers both normative and causative aspects of the program needs to be constructed.

(2) Strategies in Formulating Program Theory

After identifying what aspects of program theory are relevant, evaluators and stakeholders could go further to formulate the theory. There are three approaches for formulating program theory in general: (1) the stakeholder approach, an inductive approach that involves clarification and/or development of theories from key stakeholders for evaluation; (2) the social science approach, a deductive approach that emphasizes the use of prior theory and/or knowledge pertaining to a program for formulating program theory; and (3) the integrative approach that integrates both the stakeholder approach and social science approach in the theorizing process.

Which theorizing strategies to use may depend on how the program is designed. If an action program is based upon key stakeholders' wisdom, experience, or hunch, evaluators need to work with program designers and other key stakeholders' views in order to clarify or develop their program theory. This theorizing process usually involves reviewing related documents and having intensive interviews with stakeholders. It may be also helpful for the evaluators to visit a program site. The length of time in the theorizing process largely depends on the complexity of the program. With an adequate review of related documents and materials, a few well-organized intensive interviews with program designers and other key stakeholders may be enough for formulating program theory for a program with low complexity (Chen, 1990). However, if it is a large scale or complicated program, it could take several intensive interviews with key stakeholders and one or more site visits to formulate the program theory that reflects key stakeholder groups' views (Chen, 1997a).

If the design of a program is based upon the existing social science theory, the evaluators can apply the social science approach to formulate program theory (Chen, 1990). It is important to stress that even when social science theory is applied, it is still highly important for evaluators to work with key stakeholders in the theorizing processes. Since social science is very general, the conversion of the general social science theory into a workable program requires a lots of translation efforts and formulation from program designers and implementors. Relying mainly on social science theory and without working with stakeholders, evaluators may focus on a version of program theory from social science theory that does not reflect stakeholders' views and intentions.

(3) Finalizing the Version of Program Theory that Will Be the Focus of Evaluation

Stakeholders and evaluators need to develop a final version of program theory that will be used in evaluation. A trimming strategy is recommended for this purpose. This strategy starts from constructing a general program theory without considering the resources available for the evaluation. The general program theory can provide stakeholders a systematic view of how the program is supposed to work and what are the important assumptions for the program. In the meantime, the evaluators need to work with stakeholders in determining the priority of pressing issues needing to be answered from the evaluation. Based upon the general program theory and the priority list, evaluators and stakeholders can determine which aspects or elements of program theory are crucial and what needs to be evaluated under the existing resources available to an evaluation. If stakeholders feel it is important to evaluate a version of program theory larger than the proposed budget, this provides evaluators a solid ground for asking for more money.

APPLICATIONS OF RESEARCH METHODS: THE CONTINGENCY PERSPECTIVE

After the final version of program theory for evaluation is determined, the evaluators can base upon the program theory, the characteristics of the program, and stakeholders' needs to select a research method or methods to gather data for verifying the extent that the program theory is operated in the field. Since the nature of program theory, the characteristics of the program, and the stakeholders' needs are different from program to program, it is expected that research methods used in evaluating these programs are also different. The emphasis is on how a research method is useful for understanding or verifying program theory rather than stressing for methodological purity and loyalty. Unlike a method-driven evaluation which is bonded by a particular method or methodological tradition that is adopted in an evaluation, the framework of theory-driven evaluation allows evaluators to utilize any method that fits for their evaluation situation.

Chen (1997b) identified the contextual circumstances that are relevant to selection and application of particular research methods. Some of these contextual circumstances can be conceptualized as the following dimensions: depth vs. width information, high vs. low availability of data, and low vs. high openness of environmental influences.

Understanding the Events and Activities in Depth and in Context versus Estimating the Prevalence of the Elements in Width and in Precision

This first dimension concerns the nature of the information that is to be provided by the evaluation. For example, an evaluation may be required to provide an understanding of program events and activities in depth and in context. Stakeholders may need to know why a particular treatment or intervention was chosen, how the eligibility of the client was determined, the variety of channels that were used to reach clients, how the implementors of the program were recruited, and the feelings that both the implementors and clients had about the program experience.

The purpose of this kind of inquiry is to provide a holistic picture of the program. On the other hand, an evaluation may be required to generate precise prevalence or incidence of the various program elements and/or the relationships among them. Stakeholders may need to know how many clients actually used the services provided by the program, the social and demographic profiles of the clients, the percentages of clients that were recruited from each channel, and the impact of a treatment on the outcome after holding all other influences constant. There is also the possibility that an evaluation may be required to provide both types of information.

High Availability versus Low Availability of Credible Data

This second dimension concerns the availability or accessibility of valid and reliable data related to the evaluation. Credible data related to outcome measures such as birth rates, crime rates, highway incidents, drop-out rates, and so forth can be easily obtained from public records. In the event that few outcome measures are available in the public records, evaluators can get permission to access the target group itself for data collection. This type of data is said to have high availability. On the other hand, evaluators may face a situation where there is low availability of credible data related to a program. Program

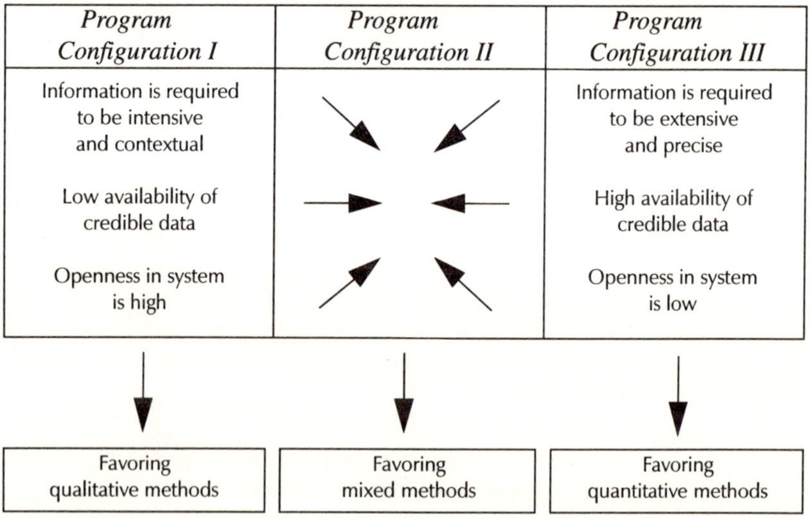

Program Configuration I	Program Configuration II	Program Configuration III
Information is required to be intensive and contextual		Information is required to be extensive and precise
Low availability of credible data		High availability of credible data
Openness in system is high		Openness in system is low
↓	↓	↓
Favoring qualitative methods	Favoring mixed methods	Favoring quantitative methods

Source: Adopted from Huey T. Chen (1997). Applying mixed methods under the framework or theory-driven evalutions. In J. Greene & V. Caracelli (Eds.), *Advanced mixed methods evaluation* (p. 61-72). San Francisco: Jossey-Bass.

Figure 2.2. Program Configurations and Choice or Methods

implementors may not trust "outside" evaluators to access the data of the program, or clients may be unwilling to reveal their feelings about the program experience to the evaluator. The situation may also arise where some credible data are readily available, but other important data are not.

Low Openness versus High Openness of Environmental Influences on a Program

This third dimension deals with the degree of influence that the environment has on the program. It can range from no influence to continuous influence, or from a totally closed to a totally open system. A program with low openness exhibits little interaction with its environment. An example of this type of closed system is a laboratory experiment which seeks to understand the effects of a drug on a certain animal. In this case, the researcher has total control of the research conditions. On the other hand, a program with high openness constantly interacts with the environment. Such a program is fluid and constantly changing. The program staff needs to continuously react and adjust to the environmental influences in order to keep the program alive, making it difficult for evaluators to control the research conditions and perform reliable measurements. Alternatively, a program may have some of the characteristics of both an open and a closed system (Chen, 1990). This type of program interacts with its environment in order to acquire resources for survival, but maintains a closed operational core in order to ensure smooth implementation and management.

Different program evaluation contexts may have different configurations of these dimensions. Often, factors of these dimensions tend to group into the following three configurations as shown in Figure 2.2.

Configuration I: This configuration, illustrated in the left-hand side of the figure, indicates those program evaluation contexts which require depth of information, have low availability of credible information, and a highly open program system.

Configuration II: This configuration, illustrated in the center of Figure 2.1, consists of those program evaluation contexts which require width or representative information, have high availability of credible information, and a fairly closed program system.

Configuration III: The right column illustration of Figure 2.1 indicates the configuration of a program evaluation context which requires information that has both depth and width, offers high access to some information but low access to other information, and has the characteristics of both open and closed systems.

These patterns suggest the following principles for selecting methods in an evaluation:

1. When an evaluation's context configuration is closer to the patterns

described in Configuration I (deep information desired, low availability of credible information, and high openness of the system), it is more appropriate to use qualitative methods.

2. When an evaluation's configuration is characterized by the requirement of wide information, high availability of credible information, and low openness of the system, as described in Configuration II, quantitative methods are the most appropriate choice.

3. When the configuration of an evaluation context most resembles that described in Configuration III (requirement for both deep and wide information, only partial availability of credible information, and characteristics of both open and closed systems), it is more appropriate to use mixed methods.

WHEN THEORY-DRIVEN EVALUATIONS ARE USEFUL OR NOT USEFUL?

Theory-driven evaluations can be applied to many evaluation situations. However, for application purposes, it is helpful to point out the conditions that favor or do not favor the applications of theory-driven evaluations. The following are the circumstances in which theory-driven evaluations are particular useful:

(1) When there is a need for a comprehensive assessment of the merits of a program.

Theory-driven evaluations are useful for those programs which are labor intensive, involve coordination among many organizations, or are complex in implementation. An assessment of this kind of program requirement information is not only the effectiveness of a program, but also the context of these results. Without contextual information, the black box type of assessment may be inadequate or even misleading. An action program is a social system in which the same program goal could be achieved through different channels. However, some of the channels may be illegitimate or not socially approved. Without knowing the exact channel that produces the effect, it is difficult to determine the merit or worthiness of a program (Chen, 1994b). For example, an education program could enhance disadvantaged students' reading scores through different channels, such as teachers providing quality tutoring to these students; teachers using physical punishments for those not performing well; or teachers giving students the tests and instructing them how to answer the questions. The first one is legitimate, while the latter two are socially disapproved or are illegitimate. Since a black box evaluation cannot differentiate which channel produces program effectiveness, it may run into the danger of claiming a program is successful, even if the program uses illegitimate or socially disapproved means to achieve its goals. As argued by Chen (1994, p. 80), in many situations, the question of "How does the program achieve the goal?" is as important as the question of "Does the program achieve the goal?" Theory-driven evaluation can provide information on these two areas in providing a balanced view of the merits of a program.

(2) When evaluation needs are to serve the development function of both the assessment and development functions.

The purpose of evaluation never limits to the assessment of the merits of a program. Another important evaluation purpose is development, that is, helping stakeholders to develop or improve their action programs (Patton, 1996; Chen, 1996). When stakeholders have a strong need to improve the quality and effectiveness of the program, theory-driven evaluations are particularly useful for serving this purpose. By examining the prescriptive and descriptive assumptions, theory-driven evaluations cannot only provide information for understanding why a program works or does not work, but also pinpoint the problematic or weakness areas or elements for future improvement.

On the other hand, there are also some conditions that do not favor the application of theory-driven evaluations. These conditions do not so much relate to the constriction of program theory as to the purpose of an evaluation. As will be discussed in the next chapter, strategies are available for facilitating evaluators to work with stakeholders in developing program theory. However, the following purposes of an evaluation can diminish the need to apply theory-driven evaluations:

(1) When Key Stakeholders are Interested Mainly in Unequivocal Evidence on the Gross Impact of the Program on Predetermined Goals.

If program stakeholders have an overriding concern about knowing whether a program has achieved a set of predetermined goals, their interest focuses on the use of the most rigorous design to produce unequivocal evidence of the gross impact on the target group of the program. Under these conditions, an evaluation uses the most rigorous design, and that is the best choice for these purposes. There is no need to do theory-driven evaluations. However, if stakeholders' interest and concern are beyond estimating gross impact such as examining program effectiveness under the context of treatment implementation, program improvement, generalizability and so on, then theory-driven evaluations are needed.

(2) When the Purpose of Evaluation is Mainly to Assess Stakeholders' Satisfaction with a Program.

If the purpose of the evaluation is mainly to provide information on the overall satisfaction of the program, stakeholders such as clients or consumers are asked questions about their satisfaction with or sentiment toward the program. Under this condition, there is very little need for doing theory-driven evaluations. However, when evaluators are also expected to provide information on the reasons or elements of satisfaction or dissatisfaction, then theory-driven evaluations are needed.

0041007

Table 2.1.   The Spring Sun Program: Normative Versus Actual

Program Domains/dimensions	Normative	Actual
Goal/Outcome	Reduction of student drug use to be verified through urinalysis	Reduction of drug use, but urinalysis collection environment not controlled
Treatment	Primary: provide quality counseling to abusers Secondary: basic drug education	Primary: counseling mainly involved use of threats, admonishment, and/or encouragement not to use Secondary: basic drug education
Implementation Environment Target Group	All drug abusing students	Only those drug abusing students who were easy to reach
Implementors	Teachers provided with adequate drug treatment training and information	Teachers lacked adequate drug treatment skills and information
Mode of delivery	Compulsory individual counseling	Compulsory individual counseling; but with problems such as lack of plan, format, and objective
Implementing organization	All schools that can adequately implement the program	Smaller schools had difficulty implementing the program
Inter-organization procedures	Effective centralized school system	Communication gap, mistrust between Ministry of Education and the schools
Micro-context	Eliminate video game arcades	Video game arcades still exist
Macro-context	Strong public support	Strong public support, but problematic education system (elitism)

Source: Adopted from: Huey T. Chen. (1997). Normative evaluation of an anti-drug abuse program. *Program Planning and Evaluation, 20* (2), 195-204.

EXAMPLES FOR DESIGNING AND CONDUCTING THEORY-DRIVEN EVALUATIONS

This section discusses three concrete research examples to facilitate the understanding of theory-driven evaluations.

(1) Normative Evaluations

The major task of a normative evaluation is to examine the congruency between the program as planned, as in normative theory, and the program that is actually implemented. Normative evaluations have long been applied in two dimensions of implementation system: target group and treatment or intervention. The focus has been on whether the treatment or intervention has been delivered as it is intended or whether the treatment or intervention reaches the intended target group (see Rossi & Freeman, 1993). However, as discussed in the previous section, the implementation system contains additional five dimensions that may be important for a particular evaluation. In fact, evaluators and stakeholders can utilize the framework to determine which domains or dimensions are to be included.

In order to illustrate the comprehensive application of normative evaluation, a large scale application of normative evaluation (Chen, 1997a) is illustrated as follows: Since 1993, the Ministry of Taiwan has implemented an anti-drug abuse program to reduce the use of drugs in the middle schools. The program was developed by a small group of top officials in the Ministry of Education. Some information relating to program operations was available in quarterly reports, but much of the relevant information needs to be gathered from the local schools. These schools were reluctant to reveal this information to outsiders. The funding agency requires that the evaluation provide comprehensive and generalizable information on how the program was operated.

Evaluators applied the stakeholder approach to develop the normative theory. More specifically, strategies such as document review and intensive interviews with program designers and local implementors were used to develop program theory. The normative theory consisted of the following domains and dimensions related to treatment, goals, and implementation of the program as represented on the left-hand side of Table 2.1.

Due to the consideration of the nature of the program, as well as the funding agency's requirements, mixed methods were applied to gather data. Because schools were highly reluctant to reveal credible information to outsiders, the evaluation team had to personally visit the schools and explain the purpose of the evaluation and the confidentiality of the data. However, the evaluation team only stayed at each school one day in order to save resources for visiting more schools in order to enhance the generalizability of the evaluation. The normative theory provides a framework for guiding and collecting data relating to how the program was actually implemented in the field. The verification of these domains and/or dimensions in the normative theory usually require a comprehensive data set gathered from the application of quantitative and qualitative methods. For example, within the implementor dimension, quantitative methods were applied to examine the prevalence of teachers' satisfaction/dissatisfaction regarding the workshop of drug abuse counseling skills sponsored by the Ministry of Education, while qualitative methods were used in probing the contextual issues such as why teachers have this kind of reaction regarding the workshop.

The empirical findings for the actual implementation of the program by the local schools are presented in the right-hand side of Table 2.1. A comparison between the normative aspects of the program and the actual implementation of the program revealed a fair amount of discrepancy between the program as viewed and planned by the Ministry of Education and the program as actually implemented in the local schools. Some of these discrepancies could be attributed to the following factors: the lack of appropriate training of teachers in drug abuse counseling skills, the high ambitions of the Ministry of Education in requiring the schools to deal simultaneously with drug abuse and other social problems, and a sense of mistrust and lack of communication between the authoritarian Ministry.

(2) Causative Evaluations

As discussed in the earlier section, causative theory is essential in the design of a causative evaluation. The time and efforts used for formulating program theory can be shortened considerably when the conceptual framework described in the last section can be used in the theorizing processes. For constructing a causative theory, the evaluators ask the question such as "Why the treatment or intervention is supposed to work?" or "What kind of channels it would require for the treatment or intervention to be effective?" Or evaluators could propose some plausible mediating or moderating factors as a starting point to work with or develop with stakeholders.

In many situations, the design of an action program is based upon program designers' and other stakeholders' ideas, hunches or experiences. Evaluators need to take the stakeholder approach to clarify or develop their theory. Since stakeholders are often overwhelmed by the terms such as theory, evaluators should avoid asking the direct questions such as "What is your causative theory?" Or, "What is your program theory?" A fruitful strategy to understand stakeholders' ideas and intentions in the intensive interview is to raise the question to program designers and other stakeholders' questions such as "Why do you think that the treatment or intervention can attain the program goals?" or, "Are there some factors that could facilitate or hinder the treatment or intervention to affect program goals?" The answer to these questions often provides useful clues for identifying the mediating or moderating factors underlying the program.

A good example for the application of the stakeholder approach to construct program theory was demonstrated by an evaluation of a school-based anti-smoking program (1988). The program designers devised a comic book with an anti-smoking story as an intervention for changing students' knowledge and attitude on smoking. Program designers' underlying causative theory began to be revealed, when evaluators probed the following question: "Why is the comic book helpful in changing student knowledge and attitudes on smoking?" Program designers' ideas started from their belief that teenagers are fond of reading and keeping comic books. Accordingly, a comic book conveying the anti-smoking messages would increase the opportunity for students to keep and read the comic book, which in turn would lead students to absorb the anti-smoking knowledge and change their attitude toward smoking.

More specifically, program designers' causative theory consists of two parts: The first part is that the comic book can attract student attention for reading the book. As mentioned earlier, this part of theory is the "action theory". The second part of the causative theory implies that the reading of the anti-smoking can lead to increase the anti-smoking knowledge and the unfavorable attitude toward smoking. This part of theory is representative to the conceptual portion of how the program is supposed to work. It is called "conceptual theory." The causative theory was integrated in the evaluation process. The result of the evaluation indicates that comic books indeed had the students' attention in keeping and reading it, but the data also indicates the keeping and reading of the comic book does not transform to attain the program goals. In other words, the action theory is successful, but the conceptual theory failed. The causative evaluation provides information not only as to whether a program has successfully attained its goals but also as to why the program leads to this kind of results. The information is very useful for future improvement of the program.

(3) Mixed Type Evaluations

The conceptual framework of theory-driven evaluations also provides a basis for interface between the normative and causative evaluations. One advantage for such linkage is that the findings of normative evaluation can be used to explain program effectiveness. An example to illustrate such possibilities has been provided by Chen and colleagues (1997) in their evaluation of a garbage reduction program. Household garbage in Taiwan had been collected by government sanitation workers on a daily basis. In order to deal with the ever-increasing amount of domestic garbage, a demonstration program was established by the Environmental Protection Agency in Taiwan to reduce the amount of domestic garbage generated by each household. The intervention strategy of the program was to enforce a policy of no garbage dumping or collection every Tuesday. Since the program was a demonstration program, the evaluation results had to be generalizable. The funding agency stressed the need to assess the effectiveness of the program, as well as to examine how well the program had been implemented, in order to provide concrete information for program improvement. Some of the data, such as daily garbage amounts in the community, were available in the records, but other information, such as data relating to program implementation, was not readily available. This program was a Configuration III type of program. In this case, both normative theory and causative theory within program theory needs to be constructed and assessed. A diagram of the theory appears in Figure 2.3.

The program theory as illustrated in Figure 2.3 provides a framework for the application of coupling mixed methods. For example, in the implementing stage, qualitative methods such as participant observation were used to deal with depth issues in the normative theory such as observing how the sanitation workers actually implement the no garbage on Tuesday policy. Quantitative methods were used to assess the prevalence issues; for example, the volume of garbage on Tuesday was reduced in the community.

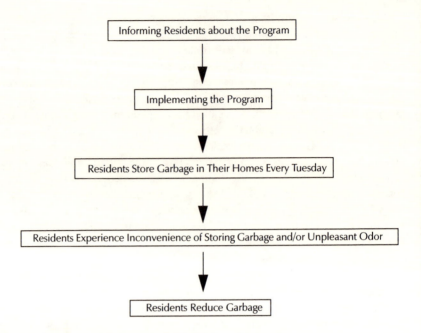

Informing Residents about the Program

Implementing the Program

Residents Store Garbage in Their Homes Every Tuesday

Residents Experience Inconvenience of Storing Garbage and/or Unpleasant Odor

Residents Reduce Garbage

Source: Adopted from Huey T. Chen, Juju Wang & Lung-ho Lin (1977). Evaluating the process and outcome
of a garbage reduction program in Taiwan. *Evaluation Review, 12* (1), 27-42.

Figure 2.3.   Casual Process in the Garbage Reduction Program

The causative theory illustrated in the bottom section of Figure 2.3 suggests
that the implementation of a policy of no garbage collection and disposal on
Tuesdays would force residents to store garbage in their homes. Because the
principal mode of residence in the highly dense demonstration city was apart-
ment housing, or a small house with no garage or yard, the inconvenience and/
or unpleasant smell of the household garbage would serve as a reminder to the
residents that the garbage problem is serious, and that it is important for residents to reduce the amount of garbage, not only on Tuesdays, but every day.

The assessment of how the normative and causative theories were actually
implemented in the community required a wide range of quantitative and qualitative data. Research staff were in the garbage disposal sites to observe how
the sanitation workers enforced the policy on Tuesdays. Residents were interviewed in terms of their reactions and experiences concerning the policy. Daily
amounts of garbage before and after the intervention were gathered. Qualitative methods were used to verify that letters were mailed, media campaigns
took place, and banners were placed on all major streets. The issue of program

effectiveness was assessed, using multiple interrupted time series design and ARIMA models.

The findings from the data indicated that the program was implemented successfully, but failed to reduce the amount of domestic garbage. This can be accounted for by the fact that the residents simply saved their garbage at home on Tuesdays and disposed of it on Wednesdays. The data from implementation indicated that the residents did not feel the inconvenience of holding the garbage at home on Tuesdays, nor did they suffer from the odor of leftover food because they tied these items in plastic bags with zippers or rubber bands.

CONCLUSIONS

The framework of theory-driven evaluation can enhance the scope and quality of evaluation as well as provide useful information for improving the programs or facilitate policy development. The advantages are summarized as follows:

(1) Providing Contextual Information about Program Implementation and Effectiveness.

By examining the crucial assumptions underlying a program, the theory-driven evacuations provide contextual information crucial to understanding the program and applying the evaluation results. For example, the traditional evaluation may focus on the issue as to whether X has an impact on Y. Theory-driven evaluations would put this question under a framework asking questions such as: What exactly is X and Y? How is the implementation system operated? Is X implemented appropriately? Is Y an appropriate goal to be evaluated? How does X affect Y? This contextual information is crucial for stakeholders to interpret and utilize evaluation results.

(2) Enhance Quality of Evidence

Because the framework of theory-driven evaluations is comprehensive for different types of research methods to make contributions, the scope and quality of evidence produced by theory-driven evaluations are broader than by method-driven evaluations. Since the focus on the theory-driven evaluations is the substantive area related to a program, theory-driven evaluations are not bounded by a particular research methodology. Pressures are imposed on different methods for making contributions on substance area related to a program rather than for stressing its methodological purity and loyalty to a particular methodological tradition. The results of applying multiple or mixed methods can strengthen the scope and quality of evidence generated in theory-driven evaluations.

(3) Sensitivity to Uniqueness of Program Characteristics and Evaluation Needs

Action programs are different in terms of stakeholders' needs and program characteristics such as maturation and environmental intrusion. The framework of theory-driven evaluations provides guidance for identifying under what circumstances, what kinds of evaluation types, principles, methods, and techniques are most appropriate. By taking such a contingency view, theory-driven evaluations have the potential to provide the evaluation that fits the best to the stakeholders' needs and enhances utilization of evaluation results.

NOTE

Chen (forthcoming) discusses strategies and techniques for constructing program theory in detail.

REFERENCES

Bickman, L. (Ed.) (1987). *Using program theory in evaluation.* San Francisco: Jossey-Bass.

Bickman, L. (Ed.) (1990). *Advances in program theory.* San Francisco: Jossey-Bass.

Chen, H. T. (1990). *Theory-driven evaluations.* Newbury Park, CA: Sage.

Chen, H.T. (1994a). Current trends and future directions in program evaluation. *Evaluation Practice* 15(3), 229-238.

Chen, H.T. (1994b). Theory-driven evaluations: If, difficulties, and options." *Evaluation Practice, 15* (1), 79-82.

Chen, H. T. (1996). A comprehensive typology for program evaluation. *Evaluation Practice 17*(2), 121-130.

Chen, H.T. (1997a). Normative Evaluation of an Anti-Drug Abuse Program. *Program Planning and Evaluation 20*(2), 195-204.

Chen, H.T. (1997b). Applying mixed methods under the framework of theory-driven evaluations. In J. Greene & V. Caraceli (Eds.) *Advancing mixed methods evaluation,* (pp. 61-72). Jossey-Bass: San Francisco.

Chen, H.T. (forthcoming). *Designing and conducting program evaluation: Beyond the basics.* Newbury Park, CA: Sage.

Chen, H.T., & Rossi, P.H. (Eds.) (1992). *Using theory to improve program and policy evaluation.* Westport, CT: Greenwood.

Chen, H.T., J. Quane, J., Garland, N. & Marcin, P. (1988). Evaluating an anti-smoking program: Diagnostics of underlying causal mechanisms." *Evaluation and Health Professions, 11*(4), 441-464.

Chen, H.T., Wang, J. & Lin, L.H. (1997a). Evaluating the process and outcome of a garbage reduction program in Taiwan. *Evaluation Review, 12*(1), 27-42.

Rossi, P.H. & Freeman, H. (1993). *Evaluation.* Newbury Park, CA: Sage.

Gramlich, E.M. & Koshel, P.P. (1975). *Educational performance contracting: An evaluation of an experiment.* Washington, DC: Brookings Institute.

Rodriguez, E., & Mead, J. P. (1997). Evaluating a community oriented primary care program: Lessons learned through a theory-oriented approach. *Evaluation and Program Planning, 20*(2), 217-224.

Shadish, W., Cook, Jr., T.D., & Leviton, L.C. *Foundations of program evaluation.* Newbury Park: Sage.

Weiss, C.H. (1997). How can theory-based evaluation make greater headway? *Evaluation Review, 21* (4), 501-524.

Weiss, C.H. (1995). Nothing as practical as good theory: Exploring theory-based evaluation for comprehensive community initiative for children and families. In J.P. Connell, A.C. Kubish, L.B. Schorr, & C.H. Weiss (Eds.), *New approaches to evaluating community initiatives: Concepts, methods, and contexts* (pp. 65-92). Washington, DC: Aspen Institute.

Wholey, J. S. (1994). Assessing the feasibility and likely usefulness of evaluation. In J. Wholey, H. Harty, & K. Newcomer (Eds.), *Handbook of practical program evaluation* (pp. 15-39). San Francisco: Jossey Bass.

Chapter 3

DESIGN ELEMENTS OF QUASI–EXPERIMENTS

William J. Corrin and Thomas D. Cook

INTRODUCTION

If an evaluator were to observe improved achievement test scores following the implementation of a new after-school academic program, she would not be able to claim with certainty that the improved scores were caused by this program. She would need to rule out all other plausible causes first. She would need to determine that the rise in scores was not due, for example, to changes in the student body, the hiring of new teachers, a change in the administration, the opening of a tutoring center across the street from the school, or other such events that could account for or contribute to the observed improvement in test performance.

In all likelihood an evaluator will not be able to determine or rule out *all* the plausible alternative explanations for the changes she observes, but the more alternatives she can rule out the greater is the certainty of claiming that a given program caused certain effects. Thus, in drawing descriptive causal inferences—that is, in answering the question: what effects has the program caused?—evaluators increase their causal certainty through a fallible process of falsification. The pro-

Advances in Educational Productivity, Volume 7, pages 35-57.
Copyright © 1998 by JAI Press Inc.
All rights of reproduction in any form reserved.
ISBN: 0-7623-0253-4

cess is fallible because undiscovered alternative causes may always exist, and because some alternatives may be ruled out spuriously, making it impossible to draw causal conclusions with absolute certainty. Nonetheless, the thorough evaluator approaches the ideal.[1]

Alternative causes vary in how difficult it is to identify them. While the various alternative causes of test score improvement mentioned in the first paragraph are all easily observable, some others are not. Thus, in a more complex design that also includes a control group, the controls may perform less well than they might have because they are disappointed or angered that they did not participate in the program (resentful demoralization) (Cook & Campbell, 1979). Or they might perform better than expected because of their competitive desire to match the performance of their counterparts who are benefiting from the program (compensatory rivalry) (Cook & Campbell, 1979; Saretsky, 1972). All alternative causes threaten the validity of a causal claim that a researcher desires to make.

Evaluators often use randomized experiments or quasi-experiments to deal with these threats. Both these approaches have a common experimental component. Each has a treatment (e.g., a test preparation program), outcome measures (e.g., achievement tests), and units of assignment (e.g., individual students, classrooms, or schools). But with randomized experiments, researchers use random assignment to assign units to treatment and non-treatment groups, a practice that creates two groups that are probabilistically equivalent to each other except for their treatment status. Any group differences observed after the treatment cannot then be a result of initial non-comparability. They are more likely due to the program under investigation. Unfortunately, in many situations it is difficult to conduct randomized experiments, and some experiments that begin with random assignment suffer from attrition as different kinds of persons drop out of a study over time, debasing the probabilistic equivalence between groups associated with random assignment. Therefore, many evaluators seek to reduce or eliminate validity threats within a quasi-experimental context where causal inferences depend on comparisons of groups that are initially nonequivalent or that become so with the passage of time. Turning to quasi-experiments, however, means that evaluators need to separate out the effects attributable to the evaluated program from those caused by the non-comparability of the various treatment groups being compared.

Quasi-experiments rely primarily upon structural design features, like pretests and control groups, to rule out alternative interpretations. This is not, though, the only alternative tradition to randomized experiments. Another tradition relies more heavily on statistical adjustments, seeking to measure alternative interpretations (or proxies for them) directly and then removing their influence from the dependent variable under analysis. There are many such adjustment techniques, best summarized in the work of Heckman (1980; Heckman & Hotz, 1988, 1989; Heckman, Hotz, & Dobos, 1987). Mathematical statisticians, however, are not very appreciative of such work, preferring instead to rule out threats by design where this is feasible and then using statistical adjustment techniques to control for any residual bias that might still be remaining after design controls have been used (Holland, 1989; Rosenbaum, 1995). However creative and relevant the structural controls used in a quasi-experiment, the reality is that some statistical analyses will nearly always be needed to control for any remaining threats of whatever size. But the central message of quasi-experimentation is to minimize the need for such theory- and measurement-based statistical adjustments through the prior use of structural design elements.

Prior presentations of quasi-experimentation have concentrated on evaluating various designs in terms of their capacity to rule out those alternative interpretations that have been called "threats to internal validity." There has been considerable confusion in the past about the meaning of internal validity. Campbell and his followers (Campbell, 1957; Campbell & Stanley, 1963; Cook & Campbell, 1979) understand it in terms of alternative interpretations to the conclusion that, in the particular circumstances of a given research project, something caused something else. In other words, because a treatment was in place something happened that would not otherwise have happened. There is no presumption here of identifying the cause or effect in general or theoretical terms; nor of exploring whether an investigator can generalize the study results to the kinds of persons or settings specified in the research questions guiding the study design. So, critics like Kruglanski and Kroy (1975) and Cronbach (1980, 1982) came to believe that Campbell's understanding of internal validity was flawed, if not trivial. Kruglanski and Kroy insisted that internal validity should include reference to the degree of confidence that should be attached to how the cause and effect should be labeled in theoretical terms, for almost all research hypotheses specify the nature of the thing that should be the cause and the nature of the thing that should be the effect. Cronbach went even further, understanding internal validity in terms of the extent to which an original research question is answered, noting that such questions typically specify not only the nature of the causal agents and impacts but also the populations of persons that should be affected and the types of situations in which the causal relationship should be manifest. Campbell (1986) resisted these extensions, but relabeled his old conception of internal validity as local molar causal validity. This emphasizes that internal validity is limited to causal relationships (hence, the causal part of the new name), to the context studied and not beyond (hence, the local component) and is associated with cause and effect operations rather than with more precise understandings of the theoretically effective treatment components, of the theoretically impacted effect components, or of any processes that mediated between the one and the other (hence, the molar part).

We retain Campbell's understanding of internal validity, as well as the language he has introduced for describing individual threats or plausible alternative causes. The threat of *history* refers to when an observed effect might be due to an event that is not the program or treatment of interest that takes place between the pretest and the posttest. *Maturation*, mentioned earlier, refers to when an observed effect might be due to the respondent's becoming stronger, older, or more experienced, or developing in some other way. *Testing* is a threat when an effect might be due to the number of times particular responses are measured (i.e., the increased familiarity of respondents with a test). If there is a change in the measuring instrument between pretest and posttest, such as when observers become more experienced during the course of an evaluation, *instrumentation* becomes a threat to causal claims about the effects of a program. *Selection* refers to when differences between the kinds of people in different experimental groups may have caused an observed effect. When an observed effect may be the result of the different kinds of people who dropped out of a specific treatment group during the course of the evaluation, causal claims are threatened by *mortality*. *Statistical regression* is a threat when a respondent performs abnormally poorly or well on a pretest measure, implying that any positive or negative change observed on a subsequent measure may not be due to the treatment in question, but rather due to the respondent's return to normal or average performance. Additional threats to the validity of causal claims include *ambiguity about the direction of causal influence*, such as when it is impossible to determine whether less teacher supervision causes greater student responsibility or greater student responsibility causes less teacher supervision; *diffusion or imitation of treatments*,[2] exemplified by control schools learning and implementing the fundamentals of a particular reform program, thus reducing or eliminating treatment-related differences from the designated reform schools, and decreasing the certainty of causal claims about the program dependent on a comparison of outcomes from treatment and control schools; *compensatory equalization of treatments*, when control groups are given another benefit (e.g., extra funding) to make up for not being able to participate in a potentially beneficial program; as well as *compensatory rivalry* and *resentful demoralization*, both defined earlier. Cook and Campbell (1979, pp. 50-59) discuss all of these threats in greater detail.

The main task of quasi-experiments is to use structural design features other than random assignment to rule out the specific threats to internal validity listed above in order to support the claim that the observed treatment-outcome relationship would not have come about in the absence of a treatment. The key here is the counterfactual: What would have happened had there been no treatment (Holland, 1989)? Of course, this is a state of affairs that cannot be fully known from comparison groups alone, especially when the treatment and control groups are nonequivalent, which is by definition the case with quasi-experiments.[3]

This chapter is unlike past work on quasi-experiments in that we concentrate on presenting the micro-design elements that facilitate causal inference rather than on presenting more elaborate designs that combine two or more design elements in trying to answer causal questions. Past presentations of such designs turned out to be useful for illustrating relative differences in the power to rule out alternative interpretations. But this type of presentation runs some dangers. One is that readers will think that past designs describe the full range of possible designs. Yet knowing many design elements enables the evaluator to create designs that have never existed before and that might nonetheless suit the needs and possibilities inherent in a particular study. A second danger is that readers will come to believe that when a design rules out a particular threat in whatever example is presented, it will then necessarily rule out this same threat under all conditions. While this is often the case, it is not invariably so. This is because the threats analysts try to rule out are, in the last analysis, a function of the local circumstances of a study as well as of the experimental design used. We do not want to belabor this point too much, for designs are nothing more than particular combinations of design elements. But the elements ultimately have more flexibility and local relevance than do designs. They function like tools in a tool kit, and the researcher can choose the appropriate elements in the appropriate combination to achieve whatever research goals are desired.

THREE RESEARCH DESIGNS

In this section we begin with the traditional way of presenting quasi-experiments—in terms of designs. This is less for the purpose of evaluating the strengths and weakness of specific designs, and more for the purpose of beginning the introduction of design elements. The researchers involved in each of the three studies we present wanted to probe causal relationships and so had a treatment group. But the combination of other design elements varies. The first study employs an untreated non-equivalent control group plus a pretest and posttest measured on the same scale; the second uses school cohorts in an effort to reduce the degree of initial non-equivalence between the treatment and control groups; and the third uses multiple pre- and post-treatment measures in order to create an interrupted time-series design.

The Untreated Control Group Design with Pretest and Posttest

Program evaluations using this basic design are sometimes causally interpretable. The use of a pretest and a posttest allows the evaluator to determine if change occurred during the period of time when the program was in place, while the use of a control group whose members do not participate in the program has the potential to test the crucial counterfactual—that is, whether the amount of observed change is greater among program participants than controls. If a convincing case can also be made *that the alternative interpretations operating on the treatment group are similar to those operating on controls*, then the evaluator's claim that the program caused the observed differential between the treatment and controls is better justified. Of course, this last assumption is problematic whenever the process of selection into treatments is not fully known.

Levinson and Felberbaum (1993) used this basic design in their evaluation of "Earn and Learn," a "goal-oriented, work experience program designed to encourage middle school students to develop a positive self-image and improve school performance" (p. 1). Middle school advisors and staff in a suburban Chicago school district identified candidates for the program based on their beliefs that those students were at risk of dropping out of school. These candidates then had to complete an application and attend an interview with their parents before finally enrolling in the program. The students participated in "Earn and Learn" during seventh and/or eighth grades, with Levinson and Felberbaum collecting data from them in the sixth (pretest) and eighth (posttest) grades. The evaluators collected data on three outcome types: academic variables including standardized achievement test data and grades for several subjects, academic motivation variables including absences and tardiness, and social behavior variables including suspensions. Data were also collected from a control group of students who were candidates for the program but did not participate either because they chose not to or because the program slots were filled. Comparison of the sixth grade data indicated that the two groups were highly similar on all the measures in each of the three outcome domains, though they were presumably different on some unmeasured variables. Unfortunately, in comparing the amount of change from the sixth to eighth grades, Levinson and Felberbaum found only one significant difference between the groups. Absenteeism increased in both groups but at a lesser rate for the "Earn and Learn" students, leading to the conclusion that "Earn and Learn" was partially but not generally effective. Hence, the evaluators called for a program redesign.

Cohort Design

Cook and Campbell (1979, p. 127) define "cohorts" as "groups of respondents who follow each other through formal or informal institutions like (siblings within) a family," or like grade levels within a school or trainees within a corporation. Cohorts are useful because some cohorts participate in programs that prior or subsequent cohorts do not; cohorts usually differ less from one another than random individuals would, making them less non-comparable as controls; and archival records are often available for providing the outcome data on which succeeding cohorts can be contrasted, such as academic performance, attendance, and behavior records that schools or school districts keep (Cook & Campbell, 1979).

It is often the case that programs implemented in some schools one academic year were not there in previous years or are discontinued in future years. This permits evaluating how a participating cohort of students compares to non-participating prior or subsequent cohorts. The crucial assumption here is that, from one year to the next, the characteristics of a school's population will remain fairly constant. Thus, there will be no changes in, say, the percentage of students who are of different racial or ethnic backgrounds, from different socio-economic strata, of different genders, or of varying academic ability. There are clearly instances when this may not be true, such as when students begin to be bussed into a school system from another community or when a community is undergoing rapid demographic changes. Evaluators must be aware of such changes that could reduce the comparability of cohorts.

Minton (1975) investigated how the first season of *Sesame Street* affected the school readiness of kindergarten children. Her research situation made the sophisticated use of a cohort design effective in her study. First of all, since *Sesame Street* was aired on television for the first time in 1970, kindergarten students from 1968 and 1969 who were never able to watch the program could be compared with 1970 students who could have watched the program. Indeed, according to data Minton collected on exposure to *Sesame Street*, the 1970 cohort did view the program a great deal. Secondly, Minton presumed that the 1968 and 1969 cohorts of students in the observed school district would be similar to those from the 1970 cohort, and subsequent comparisons of the racial and social class composition of each cohort supported this for these observed attributes (but not, of course, for unobserved ones). Lastly, the school district kept records of student scores on the Metropolitan Readiness Test (MRT), the outcome measure she used for the three cohorts.

Minton found that the scores on the Alphabet Subtest of the MRT, which tests recognition of lower-case letters, were significantly higher for the students in the 1970 cohort than for the other two cohorts. The five other subtests showed no differences. She buttressed this one positive finding with a content analysis showing that most of *Sesame Street*'s curriculum was devoted to knowledge of letters. In addition, switching to a sibling cohort design, she was able to show that the preceding effect held when the analysis was restricted to siblings from the years before and after *Sesame Street*. Overall, Minton concluded that *Sesame Street* caused the observed positive effect for alphabet recognition, and she improved her claim by using cohorts to reduce the degree of group non-comparability and by using a content analysis to indicate which dependent variables from the MRT were most likely to have been influenced by the treatment. It is difficult to think of other alternatives that would improve letter recognition but fail to influence other cognitive skills.

Interrupted Time-Series Design

Multiple observations over time on the same measure constitute a time series. The annual records a school district has of sixth grade achievement test scores by school and the records a high school has of each of its students' quarterly grade point averages are both examples of data amenable to time series analysis. If a treatment or program is introduced in the middle of this series, the task is to learn whether the program causes a change in the mean, variation, or cyclicity of the post-intervention time series relative to what went before. For example, if a student began to receive individual tutoring at the start of his junior year in high school, and if the tutoring were effective, we would expect his quarterly GPA to be higher over his junior and senior years when compared to his freshman and sophomore years.

The Education Consolidation and Improvement Act of 1981 (ECIA) changed the nature of federal support for education—from categorical funding targeted at schools with specific needs to block grants where the funds could be used more flexibly. Kearney and Kim (1990) wanted to see if the change in grant type affected local educational spending overall and the distribution of support for different types of local districts. They decided that the best way to investigate these issues was through the use of an interrupted time-series design because annual records of school district expenditures were available for 525 K-12 school districts in Michigan, and effects of the 1981 legislation could be determined by comparing spending records before and after 1981.

About their choice of research design, Kearney and Kim wrote, "The advantage of this design is that it allows for the determination of trends over time... But, in general, the quasi-experimental time-series design suffers from lack of control. For example, it is not certain that some event other that the treatment in question caused the change in the dependent variable" (p. 378). The authors attempted to deal with this uncertainty by identifying and measuring factors that co-occurred with the 1981 Act and that might have influenced spending patterns. They collected and statistically manipulated data on "changes of student enrollment, proportion of poverty children, population density, per capita income, and purchasing power" (p. 378) in order to isolate the effects of the funding change from these other plausible alternatives. Quasi-experimentalists would see these statistical adjustments as a last resort, preferring to use design strategies to rule out any threats that might have occurred at the same time as the intervention, especially a control series or a series with switching replications (Cook & Campbell, 1979). But more on that later ...

In any event, Kearney and Kim's analyses were based on the comparison of six measurements of annual expenditures per pupil prior to the treatment (the 1976-77 through the 1981-82 school years) and three afterwards (the 1982-83 through the 1984-85 school years). They observed a sharp drop in revenue in 1981-82 (a decrease of $3.27 per pupil), the year before implementation of the block grant. This premature decrease they indirectly attributed to the new provision, suggesting that education officials informally knew about the new federal policy and acted on it even before block grants were finally implemented as a legislative fact. Moreover, they also showed in sub-analyses that the block grants were less effective in providing funds to four types of districts: those that had previously received additional fiscal support from the Emergency Schools Assistance Act, those with higher proportions of poor children, those which were more highly urbanized, and those serving larger populations of students. In short, the switch to block grants had a negative effect on the funding of those poorer school districts that had benefited from the earlier sources of categorical, targeted, federal funding.

If Kearney and Kim had only compared 1981-82 and 1982-83 federal educational funding levels they would have noted a slight rise in federal expenditures per pupil ($0.52) and perhaps wrongly concluded that the switch to block grants had led to a minimal increase in federal funding. Having multiple measurements over time meant that Kearney and Kim could identify trends in funding, could observe the general drop that occurred the year prior to official implementation of block grants, and could specify the types of schools in which this drop was most likely to occur. Clearly, the type of quasi-experimental design investigators use makes a difference to the quality of the resulting causal inferences.

DESIGN ELEMENTS

Research designs are simply structures built from design elements. Among those mentioned in the previous section are: the presence and type of a control group, the way the controls are selected in order to reduce initial noncomparability, the availability and the number of pre-intervention observations, and the degree of conceptual fit between the treatment and the outcomes expected to change or not expected to change because of theoretically specified causal powers in the treatment. The three research designs discussed above represent different combinations of these design elements, and each design could have been improved by adding more such elements. Indeed, by demonstrating how different design elements could have been incorporated, it becomes clearer how evaluators can benefit from knowledge of design elements to create more novel and more locally tailored quasi-experimental designs. The eight basic design elements we want to discuss here are:

- Using control groups in general
- Using special types of controls that reduce the degree of initial group non-comparability
- Collecting pretest measures on the outcome variables
- Developing a time-series of pretest observations
- Removing the treatment after it has been in place for a period of time
- Creating replications of the treatment-outcome relationship
- Switching replications
- Using nonequivalent dependent variables, some of which should be affected by the treatment but others not.

Control Groups

A control group is an additional group of respondents who do not participate in the program being evaluated. It is a central part of the counterfactual that evaluators must try to create when random assignment is not feasible, and it tries to describe what would have happened in the treatment group had there been no intervention. Therefore the crucial issue with control groups is their degree of comparability to the treatment group. The greater the comparability, the less threatened is any causal inference. Thus, when Levinson and Felberbaum (1993) observed similar changes in both the control group and the program participants in their evaluation of "Earn and Learn," their certainty about the absence of a causal relationship was bolstered by knowing that the control group was not different from the treatment group on a host of measured pretest variables. Logically, of course, the two groups may have differed on some unmeasured variables associated with greater growth in the control group, thereby countervailing against any treatment-caused growth. This possibility becomes less plausible, the more variables there are on which pre-treatment differences fail to show up and the more relevant these variables are to theories predicting changes in the outcomes under analysis. Even so, the threat of unmeasured variables responsible for differential change rates cannot logically be ruled out. Yet consider how much less plausible a selection alternative interpretation is in this instance when compared to comparing middle schoolers in a suburban school district with middle schoolers pulled off some national data tape or middle schoolers from an inner-city urban location!

Kearney and Kim (1990) identified the lack of a control group as a drawback to their interrupted time-series design, and as the source of their need to deal with plausible alternative causal explanations through statistical manipulations of measures of these alternatives. Unfortunately, their particular research situation limited their ability to incorporate a control group into their design since the change in federal legislation they were evaluating was nationwide. Had the legislation applied to only some states or to certain kinds of schools, then other jurisdictions could have been used as controls, as would also be possible if there were independent information about which school districts complied with the new legislation and which ones did not. Any of these possibilities would have permitted testing whether a change in the time-series occurred at the time of the intervention in the treatment group but not in the control group. What we see here is a stronger design feature—the interrupted time-series—being complemented and strengthened by the addition of a control time-series. And the more similar this series is in pre-intervention behavior, the better the comparison for causal purposes.

Special Control Groups that Reduce Selection Differences: Of Stable, Matching, and Similar Cohorts

Control groups should be as comparable as possible to treatment groups, on both observed and unobserved variables that might plausibly affect the outcome. If Kearney and Kim had been able to compare Michigan school districts with those of another state, they would clearly have preferred one (or preferably more than one) control state with similar educational funding levels and similar student demographics. Statistical regression is not likely in this circumstance, since matching at the school level usually results in highly reliable school characteristics. This is not to claim that regression is impossible with aggregate measures; there are indeed some conditions where it is possible (Cook & Campbell, 1979). It is less plausible, however, because regression is in part a function of reliability and aggregate measures tend to be more reliable than individual ones. Nevertheless, some individual ones are quite reliable when measured well, including socioeconomic background variables, prior academic history, gender, ethnicity, and racial background. Thus, when Levinson and Felberbaum matched the "Earn and Learn" participants and controls, they made sure that each group contained students eligible for the program who had similar demographic characteristics and they discovered that the two groups did not differ on pretest measures of the outcome variables. Under the assumption that this no-difference finding is not due to low statistical power, this pattern of pretest values lends weight to the notion that the two groups did not differ much to start with. Although it is possible that there may be historically unique disturbances affecting one school but not the others from which comparison students were selected, or that the two groups might be maturing at different rates despite their similar starting levels, the evidence is strong that matching on stable background variables resulted in largely comparable treatment groups. Obviously, it is crucial for evaluators to describe the size and direction of pretest differences when working with formally non-equivalent groups.

When program participation is voluntary it may be impossible to match in ways that take adequate account of the difficult-to-measure characteristics that lead some people to volunteer while others do not. Matching will inevitably be partial in this situation. Where the number of program slots is too small relative to the demand, the random assignment of volunteers is then possible. In many cases programs target specific groups of subjects (e.g., students with learning disabilities, college-bound students, or urban schools), limiting the pool of volunteers to members of that target population. This was the case in Levinson and Felberbaum's evaluation of "Earn and Learn" where volunteers came from a predetermined group of eligible candidates. Even though Levinson and Felberbaum increased the comparability of the control and treatment groups by comparing students who were all candidates for "Earn and Learn," the program participants chose to accept their slots in the program while some of the controls chose not to, resulting in a selection difference based on volunteerism that might perhaps account for the greater drop in school attendance observed among controls than among program participants. Was this difference observed because students who know about a program but ultimately decide against it are different in their attachment to school than those who voluntarily choose to enroll? Nonetheless, Levinson and Felberbaum did right to match on program candidacy, for without this the selection difference would presumably have been even larger. They would have had to compare students who knew of the program, were eligible, and voluntarily chose to attend with those who might not have heard about it, might not have been eligible, and most likely could not have attended even if they had so desired.

There are other ways to increase group comparability that do not rely on matching cases and do not lead to statistical regression. For instance, cohorts can often be compared, whether they are succeeding groups of trainees in an institution, succeeding groups of students in a school, or succeeding children in a family. The guiding assumption is, of course, that cohorts reduce initial differences over what would be achieved by other means. It is not that the cohorts are identical. Indeed, we know from work by behavioral geneticists on the "nonshared environment" between siblings within a family that family life is not the same for two brothers, two sisters, or a brother and sister (Daniels, Dunn, Furstenburg, & Plomin, 1985; Daniels & Plomin, 1985; Dunn & Plomin, 1990; Rowe & Plomin, 1981). Family environments are dynamic and not static, responsive to birth order, child temperament and parent maturation, among other things. Other types of organizations are dynamic in similar ways.

Economists Currie and Thomas (1995), in their evaluation of Head Start preschool programs based on sibling cohort comparisons, suggested that the family experience of siblings might differ because of child-specific factors such as choices parents make about which child they enroll in Head Start (e.g, enrolling a child perceived as less able so that he or she can "catch up" with other siblings) or "spillover" effects such as a Head Start child teaching a non-Head Start sibling something learned in the program (the threat of treatment diffusion) or a parent doing extra for a non-Head Start child to compensate for the benefits the Head Start child is receiving (the threat of compensatory equalization). They also asserted that a family fixed-effects statistical model when used by itself to compare siblings, could over- or underestimate true program effects because it does not take these child-specific and "spillover" factors (i.e., sources of within-family differences) into account. Thus Currie and Thomas performed further statistical analyses comparing two fixed-effects models to estimate bias related to these factors. The point is that the use of cohorts imperfectly reduces selection differences between control and treatment groups and needs to be complemented by statistical adjustment procedures designed to adjust for whatever residual bias remains after design controls have been used.

Minton (1975) expected that the kindergarten students from 1968 and 1969 who were not able to watch *Sesame Street* would be similar to those from 1970 who did watch the program. Seeking to confirm this, she probed some of the cohorts' background characteristics, such as racial and socioeconomic composition, and determined that they did not differ in these attributes, though they may have differed on other unmeasured characteristics. She also examined the MRT scores of the two earlier "control" cohorts, 1968 and 1969. Again, she did not find any differences, further supporting her assumption that all three cohorts were likely comparable, though in her case the small sample sizes make reliance on accepting the null hypothesis somewhat problematic. Not content with stopping here, Minton also conducted some analyses that combined an institutional cohort feature with a sibling cohort feature. She did this by restricting some analyses to brothers and sisters who were either in the pre-*Sesame Street* cohort or in the post-*Sesame Street* cohort, showing that these two groups did not differ on five of the MRT subtests but did differ on the Alphabet Subtest that most closely corresponds with what content analysis showed was the major focus of that year's *Sesame Street* programming. In other words, the sibling cohorts were comparable in most areas of cognitive development except for that which *Sesame Street* taught most. It is difficult to come up with any simple selection alternative interpretation that could explain such a complex pattern of theoretically predicted results based on both kindergarten and sibling cohorts.

Group comparability rules out alternative interpretations based on selection. Selection has to do with any and all differences between those who do and do not participate in a program that could plausibly be related to changes in outcomes of interest. Selection can be the result of decisions made by program administrators who decide who should get what, can be the result of choices made by the participants themselves, or can arise from any other incompletely known source that separates respondents into treatment and control groups. Only random assignment and "regression-discontinuity" (Cook & Campbell, 1979; Trochim, 1984; Trochim & Cappelleri, 1992) involve fully known selection mechanisms and make it relatively easy to rule out selection. Much more common in social science practice is partially known selection, where the available knowledge about who gets what comes from acquaintance with the physical selection process and/or from examination of the ways in which background and pretest data vary between treatment and control groups. Evaluators not only need to minimize initial selection differences, but also need to know as much as possible about the selection process in order to develop better statistical models of it and to judge how plausible selection might be as an alternative interpretation of presumed treatment effects. Short of random assignment and "regression-discontinuity," nothing is better than designing the research initially so as to reduce selection by careful matching and/or the use of cohort control group designs and by direct observation of whatever selection differences might still remain. Otherwise, adjusting for selection becomes a process somewhat removed from knowledge of the selection process.

Pretest Measures on Outcome Variables

Pretests are observations on the outcome measure that are made prior to implementing a particular program. When linked to their corresponding posttests, pretests allow the evaluator to determine if change has occurred while the program was in place. Levinson and Felberbaum's evaluation depended on being able to determine if student attendance, suspensions, and test scores changed during the time the students participated in "Earn and Learn." Thus, the pretest observations were made in the sixth grade prior to the program and the posttest observations were made in the eighth grade, after it. Determination of change in some form or another is necessary for inferring causal relationships, but it is clearly not sufficient. Indeed, Levinson and Felberbaum discovered changes among the "Earn and Learn" students, but these were generally similar to those observed among the controls.

Consider, however, the costs of not having a pretest, given that in nearly all circumstances the pretest will correlate more highly with a later measure of itself (viz., the posttest) than will any combination of other variables measured at the pretest time. This means that the pretest is superior in its functions as (1) an indicator of the extent of selection; (2) a means to increase statistical power; (3) a control variable in statistical adjustments for selection; and (4) a component of the most intuitive understanding of "change." It is not that the pretest is infallible in these functions; nor that it cannot be supplemented by additional information that will index selection and control for error and bias due to selection. It is merely to note that the pretest is by far the best single piece of information for helping meet the four goals listed above. It is a crucial element of nearly all quasi-experiments. In theory, it is less necessary for randomized experiments. In practice, it is not needed in experiments that take place in highly controlled settings where there is little fear of attrition (and differential attrition) from the study. In such tightly controlled contexts, pretests are not so crucial, for observed posttest differences cannot be due to pretest differences that the random assignment procedure has ruled out. In open-system contexts the situation is quite different, and the likelihood of differential attrition should lead all evaluators to collect pretest information for later use in describing and explaining selection problems.

Time-Series Observations (Repeated Measures)

An evaluator who employs the same pretest measure at many different time points prior to an intervention has conducted a time-series. She then has the chance to examine empirically how the series was changing over time so as to probe whether the posttest observations differ from the pretest trend or any other attribute of change over time, like the pattern of variability. This rules out many threats to internal validity. It also allows the investigator to see whether there is some unusual change in the series just prior to the treatment that might itself be a cause of the treatment being administered at that particular time and hence a potential source of statistical regression. The evaluator using a time-series methodology has to be careful here for some inflections just before the intervention can have substantive meaning when there is a viable theory of anticipatory change, meaning a change in the outcome that comes about in anticipation of a new treatment or intervention—the very interpretation Kearney and Kim offered for their results where the expected outcome change occurred a year prematurely. Such premature changes are thought to be common when studying new legislation because the merits of a proposed change are broadly debated among interested parties before the legislation is actually passed. Thus, the change itself is hardly "news," though the pre-change deliberations can definitely be so.

It is very important with time-series to get detailed information about the circumstances immediately before, during, and after the treatment. This is not just to make accurate decisions about implementation activities and hence to pin-point the intervention onset. It is also to observe the vigor of implementation, for it is often the case that a new idea slowly disseminates through a population. In this case, a time-series of the *implementation* process would not show an abrupt onset, a steep step function, and it would then be unrealistic to expect that the *outcome time-series* will suddenly change when the treatment is first introduced. In some senses, the outcome time-series has to track the program implementation series. If the latter is delayed or slow to diffuse through a population, then it is unrealistic to expect large impacts at the time an intervention is officially introduced.

Hence, the implementation of an educational program should be as rapid, efficient, and carefully executed as possible (Marcantonio & Cook, 1994). A program must be implemented attentively and efficiently to insure that it is carried out faithfully. High quality implementation means that the evaluator is making observations about a program whose form in the field closely matches its form in theory. Sloppy, careless, or slow implementation will reduce the chances of observing treatment effects and will leave readers of evaluation reports with a quandary: Are the disappointing outcomes due to an incorrect substantive program theory, or to poor quality implementation, or to a poorly designed evaluation? It is always a responsibility of the evaluator to know the program theory well and to carefully monitor its administration.

From an internal validity standpoint, the major difficulty with a simple interrupted time-series is that it cannot account for history—forces other than the treatment that might affect the outcome and that occur, or somehow change in intensity, at about the same time the treatment is introduced. Their temporal co-variation with the treatment makes causal conclusions problematic. Several ways exist to deal with history. One is to study in great detail the period around the intervention to see what else changed with it. Another, and perhaps better way, is to obtain a no-treatment control time-series from some other, matched place. A third is to get a time-series of some outcome other than the one in the causal proposition under test, an outcome that should be affected by most plausible alternative interpretations. Thus, in a study of the effects of the British breathalyser, Ross, Campbell, and Glass (1970) were able to show that the breathalyser initially reduced serious traffic accidents during the hours when pubs were open or had just been closed but did not influence traffic accidents at other times when there was presumably less drinking but when other causes of traffic accidents were presumably the same—for example weather differences or changes in the safety features on cars. Weather and safety features can change from year to year in an annual time-series, but they are not likely to vary by time of day within the same year!

Time-series are also useful for identifying delayed causation. Given delays in implementing many policy changes it is unrealistic to expect large impacts at the time of onset of the treatment. If the implementation is closely monitored, the temporal pattern of implementation will provide clues as to when effects might be expected. Without this, causal inference is more difficult because many more history alternative interpretations are plausible the longer is the gap between treatment onset and the appearance of a possible effect. Even so, it is worthwhile looking for delayed effects, a precondition for which is that there be multiple observations after the treatment. In this connection we should note the risk that Kearney and Kim ran in having so few post-treatment observations, though this was forced on them by the need for rapid feedback on such a fundamental federal policy change. The reality is that Michigan could have changed its own local policies to counteract the anticipated effects of the federal policy, perhaps setting up targeted programs of its own, the implementation and effects of which could take a few years to appear and which would be lost to a short time-series. These are only speculations, of course, the major point being that more complete and more finely nuanced stories about the effects of policy changes depend, not just on a long pre-treatment time series and not just on monitoring implementation of the treatment, but also on the availability of a long post-treatment time series.

Treatment Removal

When the research situation makes it impossible to establish a control group, there are some partial solutions that are designed to approximate the presence of a control group. One of these involves treatment removal following treatment introduction. With measures taken before and after the treatment, an estimate of the treatment effect can be made that is, frankly, by itself very weak. Later removal of the treatment provides the potential to examine whether the opposite effect occurs after treatment removal when compared to what happened before removal. Many threats to internal validity will not be able to explain why it is that a treatment effect comes at one time and then diminishes at another time, with the times being closely correlated with the presence or absence of the treatment. Essentially, treatment removal serves to create the counterfactual baseline that is usually provided by a control group. Thus, if Kearney and Kim had control of the federal legislature and repealed the ECIA, they would have been able to observe whether educational spending patterns changed in one direction after 1981 but in the opposite direction after the legislation was repealed.

Treatment removal is not always feasible. Not all programs are designed in such a way that they can be "removed." For example, even if a school stops offering a study skills course, the students participating in it may have absorbed or learned those skills and continue to use them, showing no change in academic performance after the course has stopped. Most programs are not under the control of the evaluator, and it is unrealistic to expect that Kearney and Kim could have reversed the federal legislation or that Minton could have reversed the availability of *Sesame Street*. Also, there may be ethical problems with deliberate treatment removal, particularly if the treatment was meant to have or was exhibiting ameliorative effects.

However, treatment removal does sometimes spontaneously occur and then it is often possible to take advantage of it, provided that the reason for removing the treatment is not that the original effect has already disappeared. In such a case, there would be no way for the outcome to revert to the pretest level. Also, removing the treatment can sometimes lead to resentful demoralization or compensatory rivalry because individuals enjoyed a benefit that was then taken away from them. By itself, treatment removal is not a strong reed around which to build a cause-probing study, but it can serve as a useful adjunct in studies with other and individually stronger design features.

Replication

In some evaluations it may be possible to implement a treatment repeatedly. The simplest case of this is when there is a single sample and the treatment is repeatedly introduced and removed, as in the classical Skinnerian designs of experimental psychology. An obvious difficulty with this is that it is not feasible wherever a program is expected to have long-term effects, for these will not have faded when the later applications of the treatment are put into place.

A second case of replication is when there are multiple samples and the intervention begins at one time with one sample, at another (known) time with another sample, and at another (known) time with a third sample, and so on. The key here is to dissociate the treatment from a particular historical time and to eliminate the need for treatment removal. If the same pretest/posttest differences were observed each time the program was introduced, this would increase the certainty of an evaluator that the program caused particular effects. It would not be definitive, of course, since a simple maturation alternative interpretation would also be viable, as would a statistical regression interpretation if the treatment were only introduced because individuals seemed to be doing particularly poorly or well relative to their own past. This is why the preferred form of replication is linked to the presence of something functioning as a control group to provide a version of a counterfactual baseline. Switching replications does this.

Switching Replications

Switching replications involves reversing the roles of the treatment and control groups. In the first phase, a program is implemented with one group while the other serves as controls. In the second phase, the original controls receive the treatment and the original treatment group gets no further treatment or might even, in some special cases, have the treatment removed from it. Thus, the original controls become the replication and the original treatment group becomes the later controls. The observation of similar program outcomes during each phase strengthens the evaluator's causal attribution, just as the observation of contradictory effects would weaken it. It might suggest the need for an investigation of the original treatment group in particular, for the interpretation of outcomes depends heavily on there not being any kind of long-term treatment effect that precludes registering history-based changes in the original treatment group. The most problematic outcome is, of course, if the treatment has a long-term, growing impact, for the growth will make it unfit as a counterfactual baseline in the second phase. But impacts that increase with time are not the norm in program evaluation. Much more common are modest initial effects that then slowly dissipate. Under this condition the use of switching replications is more viable, as it is even if the initial effect persisted totally, provided that the scale position of the effect is not so high as to preclude registering true history effects that occur in the last phase.

An added benefit of incorporating switching replications into an evaluation is that it addresses potential problems that could result if a potentially beneficial program has to be withheld from some individuals, perhaps because it is in scarce supply. With switching replications, control groups can participate in the program, although at a later time. Knowing about the possibility of future involvement in the program may reduce the resentful demoralization and compensatory rivalry that can sometimes occur with members of no-treatment control groups.

Nonequivalent Dependent Variables

Nonequivalent dependent variables are outcome measures that, in theory, are not sensitive the causal forces through which the treatment is to exercise its influence but are sensitive to all or most of the alternative causal forces that might lead to spurious treatment effects. In her evaluation of the effects of *Sesame Street* on the school readiness of kindergarten children, Minton examined the scores for each of the six subtests of the MRT, knowing that *Sesame Street* spent much more time teaching letter recognition than anything else. When she discovered that scores on the Alphabet Subtest increased, but did not change on other subtests, this allowed her to construct the argument that alternative interpretations based on, say, parents working on readiness skills with their children at home, should have affected some of the other five subtest areas, such as Word Meaning or Matching and not just Alphabet Recognition. Thus, having nonequivalent dependent variables contributed to the certainty with which Minton was able to draw causal inference about *Sesame Street*. She was able to draw a more specific causal conclusion than she had begun with: it is not that the program increases all school readiness skills, it only increases alphabet recognition.

Table 3.1. Summary of Design Elements

Design Elements	Key Features
Control groups: *Groups of respondents who do not participate in the program under evaluation*	• Approximates what might have happened had the program not been implemented (the counterfactual) • The more comparable treatment and control groups are, the less threatened are causal inferences drawn from the data
Special control groups that reduce selection differences: *Groups of program non-participants selected to increase comparability with the treatment group(s)—e.g. matched groups or cohorts*	• The greater comparability between treatment and special control groups increases causal certainty by lessening the threat of selection differences as alternative causal explanations
Pretest measures on outcome variables: *Observations made prior to program implementation*	• Crucial component of almost all quasi-experiments • Necessary as a point of comparison to determine "change," an essential condition for drawing causal inference • Functions as an indicator of the extent of selection, a means to increase statistical power, or a control variable in statistical adjustments for selection
Time-series observations (repeated measures): *Employing the same pretest measure at many different time points*	• Necessary for the determination of trends or patterns among outcomes, thus increasing the evaluator's causal certainty since effects observed to deviate from the trend are more likely to have been caused by the program
Treatment removal: *Removal of a treatment after its introduction (terminating an implemented program)*	• Approximates the presence of a control group, creating a counterfactual baseline—when treatment is removed, do observed effects reverse or disappear? • Does little to improve causal certainty unless used in conjunction with other design elements

(continued)

Table 3.1.   (Continued)

Design Elements	Key Features
Replication: *Repeated introduction and removal of treatment with one sample* —or— *the introduction of treatment to one sample after another at progressively later times*	• Simulates the use of multiple treatment and control groups • Increases causal certainty if similar effects are observed with each implementation of the program • Not feasible with a single sample if the program is expected to have long-term effects • Problematic with multiple samples if each sample is subject to different circumstances or historical conditions
Switching replications: *Reversal of the roles of treatment and control groups—one group is the control while the other receives treatment, then the control group receives treatment while the original treatment group receives no further treatment or has treatment removed*	• Adds the counterfactual to the *replication* element • Makes it possible for all experimental groups to benefit from a potentially ameliorative program, reducing the threats of compensatory rivalry and resentful demoralization • Not feasible if the program is expected to have long-term effects
Nonequivalent dependent variables: *Outcome measures that are not sensitive to the causal forces of the treatment, but are sensitive to all or most of the alternative causal forces that might lead to false treatment effects*	• Effective in ruling out alternative causal explanations closely related to but not specific to the program

Evaluators can intentionally include nonequivalent dependent variables. For instance, in evaluating a program designed to improve math achievement it might also be possible to include reading scores, for the latter are sensitive to the many forces that increase achievement generally. Thus, they can serve as controls for the effects of these forces, whether they are more tutoring programs, more homework help from parents, or smaller class sizes. The key, though, is that many or all of the plausible internal validity threats will affect the nonequivalent dependent variables while the treatment will affect only the outcomes of interest in the causal hypothesis under study.

CONCLUSION

Causal description is a central goal for many program evaluators. They can therefore benefit from understanding the various design elements out of which practical designs are constructed so as to maximize their relevance to the constraints and opportunities of specific research contexts. This is important, for threats emerge from concretely embedded settings in which people and institutions live and grow. There can be no claim that a specific quasi-experimental design (that is, a specific combination of design elements) invariably rules out a particular set of internal validity threats. It may often or usually achieve this, but not invariably. Thus, we have tried to present the major design elements, other than fully known treatment assignment, out of which each evaluator can build his or her own design to suit his or her own circumstances. In a sense, we have tried to explicate the components and process out of which designs are made.

Evaluators, however, must recognize that as beneficial as design elements can be in probing causal descriptive issues, they are complemented by both substantive theory and statistical adjustments. Substantive theory is crucial in evaluation for beginning the process of exploring why it is that a particular pattern of results has come about. Design elements are crucial for fostering inferences about program effects, given the current limitations of statistical adjustments for selection if a randomized experiment is not feasible. Statistical adjustments remain necessary for dealing with those internal validity threats that structural design controls were not able to address. Evaluation needs all three techniques, and usually in the same research project. Thus, if we can achieve a study with strong design controls through the design elements described here, then our causal inference is all the stronger. If we can address any remaining issues about the quality of causal inference with the best available (but still imperfect) statistical adjustment techniques, inference is even stronger. And if analysis of mediating processes indicate that those specified in the program theory are plausible contenders as carriers of causal inference from program inputs to individual outcomes, then we should be all the more confident that the program is effective and for the reasons specified in its theory. We are not suggesting that these three methods address the same kind of question or represent equally valid methods for answering simple descriptive causal questions, but rather that each has a complementary and supportive role to play in evaluation.

Not only do design elements play an important role in program evaluation, but they are also beneficial to social science research in general. Design elements make it possible at the outset of research to rule out alternative explanations for observed relationships between data; and they also limit the dependence on statistical adjustments to control for validity threats when these adjustments have to be based on (potentially flawed) substantive theory, on (potentially invalid) measures of threats, and on (often underprobed) statistical adjustment techniques. They are the building blocks for a design-centered approach to ruling out alternative interpretations. We believe that, in the long term, social science studies that result in causal conclusions that are more likely to be true will increase the credibility and utility that policymakers attribute to evaluation and other social science findings.

NOTES

1.   Popper (1959) argues for deduction over induction since theories cannot be logically conclusively confirmed but only logically rejected. His falsificationist approach implies that theories can only be provisionally accepted if they cannot be disproved, because there may always exist a case in which they do not hold true. Our support of ruling out plausible alternative causes is based on this position of Popper's. However, it is important to understand the important criticisms of Kuhn (1962, 1976), who points out two flaws to Popper's position: the "theory-ladenness of observations" (researchers make observations colored by the theoretical lenses through which they see the world) and the incommensurability of theories (the major premises of theories are not fixed or measurable, although researchers endeavor to measure them when collecting and analyzing data). As effective as the comparison of competing causal explanations is for improving causal certainty, it is still a fallible process which must be performed with an awareness of its imperfections.

2.   These last threats are more aptly called threats to construct validity (Cronbach, 1982), but for simplicity of exposition only we include them here.

3.   Using this description of quasi-experimentation, we can see why compensatory rivalry and compensatory equalization are not threats to internal validity. They would not have occurred unless a treatment were in place, while threats to internal validity involve forces that would affect the outcome even if there were no treatment at all.

REFERENCES

Campbell, D.T. (1957). Factors relevant to the validity of experiments in social settings. *Psychological Bulletin, 54,* 297-312.

Campbell, D.T. (1986). Relabeling internal and external validity for applied social scientists. *New Directions for Program Evaluation, 31* (fall), 67-77.

Campbell, D.T., & Stanley, J.C. (1963). Experimental and quasi-experimental designs for research on teaching. In N.L. Gage (Ed.), *Handbook of research on teaching.* Chicago: Rand McNally.

Cook, T.D., & Campbell, D.T. (1979). *Quasi-experimentation: Design & analysis issues for field settings.* Chicago: Rand McNally Publishing Company.

Cronbach, L.J. (1980). *Towards reform of program evaluation: Aims, methods and institutional arrangements.* San Francisco, CA: Jossey-Bass.

Cronbach, L.J. (1982). *Designing evaluations of educational and social programs.* San Francisco, CA: Jossey-Bass.

Currie, J., & Thomas, D. (1995). Does Head Start make a difference? *The American Economic Review, 85*(3), 341-364.

Daniels, D., Dunn, J., Furstenburg, F.F. Jr., & Plomin, R. (1985). Environmental differences within the family and adjustment differences within pairs of adolescent siblings. *Child Development, 56,* 764-774.

Daniels, D., & Plomin, R. (1985). Differential experience of siblings in the same family. *Developmental Psychology, 21*(5), 747-760.

Dunn, J., & Plomin, R. (1990). *Separate lives: Why siblings are so different.* New York: Basic Books.

Heckman, J.J., (1980). Sample selection bias as a specification error. In E.W. Stromsdorfer & G. Farkas (Eds.), *Evaluation Studies Review Annual*, (Vol. 5, pp 13-31). Newbury Park, CA: Sage Publications.

Heckman, J.J., & Hotz, V.J. (1988). Are classical experiments necessary for evaluating the impact of manpower training programs? A critical assessment. *Industrial Relations Research Association 40th Annual Proceedings*, 291-302.

Heckman, J.J., & Hotz, V.J. (1989). Choosing among alternative nonexperimental methods for estimating the impact of social programs: The case of manpower training. *Journal of the American Statistical Association, 84*, 862-874.

Heckman, J.J., Hotz, V.J., & Dobos, M. (1987). Do we need experimental data to evaluate the impact of manpower training on earnings? *Evaluation Review 11*, 395-427.

Holland, P.W. (1989). Comment: It's very clear. *Journal of the American Statistical Association, 84*, 875-877.

Kearney, C.P., & Kim, T. (1990). Fiscal impacts and redistributive effects of the New Federalism on Michigan school districts. *Educational Evaluation and Policy Analysis, 12*(4), 375-387.

Kruglanski, A.W., & Kroy, M. (1975). Outcome validity in experimental research: A reconceptualization. *Journal of Representative Research in Social Psychology, 7*, 168-178.

Kuhn, T.S. (1962). *The structure of scientific revolutions.* Chicago: University of Chicago Press.

Kuhn, T.S. (1976). Theory-change as structure-change: Comments on the Sneed formalism. *Erkenntnis, 10*, 179-199.

Levinson, J.L., & Felberbaum, L. (1993). *Work experience programs for at-risk adolescents: A comprehensive evaluation of "Earn and Learn."* Presented at the Annual Meetings of the American Educational Research Association (Atlanta, GA).

Marcantonio, R.J., & Cook, T.D. (1994). Convincing quasi-experiments: The interrupted time series and regression-discontinuity designs. In J.S. Wholey, H.P. Hatry, & K.E. Newcomer (Eds.), *Handbook of practical program evaluation.* San Francisco: Jossey-Bass.

Minton, J.H. (1975). The impact of *Sesame Street* on readiness. *Sociology of Education, 48*, 141-151.

Popper, K.R. (1959). *The logic of scientific discovery.* New York: Basic Books. (Originally *Die Logik der Forschung*, 1935.)

Rosenbaum, P.R. (1995). *Observational studies.* New York: Springer-Verlag.

Ross, H.L., Campbell, D.T., & Glass, G.V. (1970). Determining the social effects of a legal reform: The British "breathalyser" crackdown of 1967. *American Behavioral Scientist, 13*, 493-509.

Rowe, D.C., & Plomin, R. (1981). The importance of nonshared (E_1) environmental influences in behavioral development. *Developmental Psychology, 17*(5), 517-531.

Saretsky, G. (1972). The OEO P.C. experiment and the John Henry effect. *Phi Delta Kappan, 53*, 579-81.

Trochim, W.M.K. (1984). *Research design for program evaluation: The regression-discontinuity approach.* Newbury Park, CA: Sage Publications.

Trochim, W.M.K., & Cappelleri, J.C. (1992). Cutoff assignment strategies for enhancing randomized clinical trials. *Controlled Clinical Trials, 13*, 190-212.

Chapter 4

CROSS-DESIGN SYNTHESIS

Robert F. Boruch and George Terhanian

This paper focuses on the *design* of field studies on which productivity research is based. The emphasis is on exploiting a new approach to analysis, notably cross-design synthesis, to facilitate the joint planning and use of controlled experiments and of passive surveys.

In what follows, cross-design synthesis and related matters are defined. Assumptions are made plain. They pertain mainly to federal institutional policy for generating the data that sustain productivity research. We lay out a rationale for the approach based on scientific and political/institutional standards.

There are two illustrations. The first concerns literacy programs in the United States. The second bears on grouping students in the schools. Both exploit contemporary evidence, of course. And both illustrate how the idea of cross-design synthesis might be depended on to improve research on production functions in this country and others.

CROSS-DESIGN SYNTHESIS

Cross-design synthesis is a strategy for combining analyses of the data that are generated in controlled experiments with analyses of data generated from surveys or administrative data bases. For example, the data obtained in a National Center

Advances in Educational Productivity, Volume 7, pages 59-85.
Copyright © 1998 by JAI Press Inc.
All rights of reproduction in any form reserved.
ISBN: 0-7623-0253-4

for Education Statistics (NCES) national probability sample survey on adult literacy in the United States might be used in a model-based analysis that purports to yield estimates of the relative effects of certain literacy programs. These survey results would then be combined, in a cross-design synthesis, with evidence generated by a small number of controlled randomized experiments on the relative effectiveness of literacy programs.

The object of cross-design synthesis is to produce valid *and* generalizable estimates of the relative effectiveness of programs. The rationale is that the combination exploits a benefit of controlled tests, notably an unbiased estimate of the treatment effect in local settings. And it exploits a benefit of national probability sample surveys, that is, the capacity to generalize to a larger target population. Roughly speaking, the same distinction was made by Walberg (1993). He described how structural models might be used to get beyond the limitations of field experiments and simple linear model-based analyses of education production functions. Our object here is to direct attention to the design of data collection efforts, rather than to analysis.

In the adult literacy case, for example, controlled experiments in particular sites may yield statistically unbiased estimates of the relative effectiveness of different literacy programs. But the estimates are usually local, for example, of uncertain generalizability. The national data base or survey may yield estimates of the effect of programs at the national level. But these estimates are usually suspect in that their validity, that is, statistical unbiasedness, is unclear because the survey or administrative data base involves no active control. Rather, analysis of such data usually involves only statistical control and depends heavily on the accuracy of the underlying models. A combination of the two sources of evidence, controlled experiments and passive surveys, might be combined so as to justify inferences that are both valid and generalizable.

The general approach to cross-design synthesis is described in a U.S. General Accounting Office report (USGAO, 1992) and in Droitcour, Silberman, and Chelimsky (1993). A more recent report (USGAO, 1995) describes the approach's application to the problem of estimating the effect of breast conservation versus mastectomy on the five-year survival rates of women with breast cancer. This analysis is based on data from randomized clinical trials and a large data base. In particular, six studies serve as the evidence in the randomized trial category; they include single-site and multi-site experiments undertaken in North America and Europe. The National Cancer Institute's Surveillance, Epidemiology, and End Results (SEER) system constitutes the administrative data base. It provides data on breast cancer patients, their treatment, and prognosis based on reports from practicing physicians in a large geographic region of the United States.

An interesting variant on the idea of jointly designing passive data bases and randomized trials has emerged in the United Kingdom (Engels & Spitz, 1997; Charlton, Taylor, & Procter, 1997; Procter, Taylor, Stark et al., 1995). Called the Population Adjusted Clinical Epidemiology (PACE) strategy, its purpose is to reduce the limitations of large scale clinical trials, notably with respect to under-standing and enhancing their generalizability, by depending on an epidemiological surveillance system that itself depends on data from populations that are not screened so narrowly as those in the trials. In this particular respect, the aims of cross-design synthesis and PACE are the same.

Although these methods of combining evidence from different sources and their application are innovative, the underlying thinking is not. Boruch (1975), for instance, framed the problem in terms of routine coupling of design of experi-ments to ex-ante design of quasi-experiments. Fienberg and Tanur (1987) consid-ered related issues in terms of "embedding" controlled experiments in larger passive studies. Still other treatments of the topic can be handled as variants on "satellite policy." Just as NASA invites special studies to be undertaken in its sat-ellite, a survey or surveillance organization may invite special studies, notably controlled experiments, to capitalize on the massive survey vehicle, that is, the sat-ellite (Boruch & Terhanian, 1996).

In what follows, we depend on the cross-design synthesis work to guide study design. Fuller explanation of the approach is integrated with the illustrations below.

OBJECTIVE, ASSUMPTIONS, AND DEFINITIONS

Objective

Recent reports on cross-design synthesis have focused on the *analysis* of data from two kinds of sources: controlled experiments, and passive surveys or/and data bases. Here, the focus is on how the thinking about cross-design synthesis can improve the *design* of surveys or administrative data bases and national surveys sponsored by NCES and other statistical agencies, and the *design* of controlled field experiments sponsored by other policy-oriented government agencies.

To put the objective bluntly, we want to turn "cross-design synthesis" into a vehicle for better design of studies, rather than to be content with its current use as a special form of meta-analysis. This objective accords with the broad theme of educing the implications of new analytic approaches for better study design. It is also distinctive; the inventors of cross-design synthesis did not develop this impli-cation (Droitcour & Chelimsky, 1995; Boruch, 1995).

Assumptions

A first assumption here is that it is important to understand what influences edu-cational productivity in the United States. Second, we assume that, as in science more generally, the development of theory of education production functions will rely on studies of different kinds sponsored by different foundations and agencies.

Institutional independence and fragmentation often underlie methodological independence and diversity. Third, we assume that, at times, the design of studies can exploit the diversity so as to improve a jointly dependent product, for example, more accurate production functions. Finally, we assume that, in five years, we will have to combine results from different sources so as to reach better conclusions about productivity.

Definitions

A controlled field experiment is a setting in which individuals (or other entities) are assigned to program variations in accord with a plan designed to produce an unbiased estimate of the differences among the program variations and a statistical statement bearing on one's certainty about the results. For instance, one may design a study to compare certain approaches to teaching English as a second language so as to understand which approach works best, and under what conditions. Individuals or entire organizations might then be randomly assigned to the different program approaches, engaged in the relevant approach, and then measured with respect to their English proficiency. The focus here is on randomized controlled experiments.[1]

Such experiments have been sponsored periodically by states, private foundations, and federal agencies. The Planning and Evaluation Service of the U.S. Department of Education, for instance, is mandated to do evaluations; a few of these are controlled experiments. The Department's Office of Educational Research and Improvement (OERI) is mandated to sponsor research, including experiments. A handful of private foundations have supported high quality social research including controlled experiments. The William T. Grant Foundation and the Ford Foundation are among them. Each has attempted to leverage larger federal efforts in related areas.

Because controlled experiments are difficult to mount, only a few are usually undertaken in a small number of sites. The results may be relatively unequivocal in the sense that one program variation appears to work better than another in one or more of the sites. It will usually not be clear how these results can be generalized. For instance, the experiment sites may include U.S. cities in the northeast. They then exclude the northwest and southwest where culture, economic contexts, and other characteristics may influence the magnitude of "effect sizes."

Survey here means an effort to elicit information from a probability sample of individuals or institutions who are members of a well-defined target population. Such a survey involves no active treatment or manipulation of respondents, apart from the act of eliciting information. The survey may be cross-sectional, for example, the 1991 National Adult Literacy Survey. Or, the survey may be longitudinal, as in the case of the National Educational Longitudinal Study undertaken in 1988 (NELS, 1988).

Administrative data base or surveillance and reporting system here is defined as a set of records on a well-defined target population. For instance, transcripts on all students in a junior college, containing information about the students' courses and grades, constitute a data base. The records on all students in a voluntary service organization's program on literacy also constitute a "data base." A data base can be regarded as a "survey" to the extent that such administrative records are the product of interviews that are also done in survey research, albeit under different conditions.

In this context, it seems sensible to depend on surveys and data bases sponsored by the National Center for Education Statistics (USDE, 1995). NCES surveys provide data used in many analyses of education productivity (Reynolds & Walberg, 1993).

THE SCIENTIFIC RATIONALE FOR CROSS-DESIGN SYNTHESIS

Users of surveys produced by NCES and others have often employed the data in estimating the relative importance of various factors, including specific programs, on educational productivity (e.g., Terhanian, 1996). It is reasonable to expect these efforts to continue despite the ambiguity in the interpretation of the data that is bound to occur because the survey is a passive instrument rather than an active experiment. Insofar as cross-design synthesis carries a promise to combine such survey data with other data from experiments, so as to produce better information, it is sensible for NCES, among others, and for policy researchers to exploit opportunities presented by cross-design synthesis.

When experiments and surveys have been designed and executed jointly, the products will be useful even when resultant data *cannot* be combined in a cross-design synthesis. Failure may occur for instance if, despite planning, no trustworthy propensity scores can be constructed so as to provide adequate estimates of productivity parameters (more on this below). Nonetheless, this scenario permits accrual of reliable knowledge and fosters theory about when alternatives to randomized experiments suffice and when they do not. That the methodological features of alternatives are crucial is clear from contemporary methodological research. Meta-analyses by Wilson and Lipsey, for instance, suggests that about 25 percent of the variance in estimated effect sizes from nearly 400 studies is attributable to features of study design and execution (Wilson, 1995).

Although controlled experiments that bear on productivity are not in ample supply, many can be readily identified. Moreover, they can be fit to theoretical frameworks developed from analyses of mostly passive surveys. These frameworks include, for instance, Reynolds and Walberg's (1993) structural models of nine classes of factors and a five stage theory of productivity. They include related meta-analytic work by Greenwald, Hedges, and Laine (1996) and Hanushek (1989).

Instructional quality, for instance, appears in such theories. This may be construed as being driven partly by environment, including school or class size. Finn and Achilles' (1990) remarkable state-wide experiments on class size exemplify a relevant genre of controlled randomized studies. The driving factors is such theories also include teacher characteristics. Some controlled experiments have been mounted, successfully and otherwise, to discern effects of various teacher-related factors, for example, Bickman (1985). Independent of these, background characteristics of students and family surface in most theory. Efforts to actively enhance motivation and ability of both children *and* their mothers have been studies through multi-site experiments on Even Start (St. Pierre et al., 1996), among others. Similarly, the U.S. Department of Education's evaluation of Upward Bound programs for enhancing the achievement of at-risk youth in schools involves nearly 70 randomized experiments (Myers & Schirm, 1997). Macro-level experiments designed to alter the environment at the school level, climate being theoretically important, have also been mounted, most often to prevent high risk behavior rather than to enhance productivity directly. They have been executed, using schools as the unit of random assignment, by among others Ellickson and Bell (1990). Other illustrations are identified below.

POLITICAL/INSTITUTIONAL RATIONALE FOR INVERTING CROSS-DESIGN SYNTHESIS

A reason for inverting the cross-design synthesis idea so as to focus on design of surveys and experiments is that the approach can be a bridge between the members of the federal statistics agencies on the one hand and the federal evaluative agencies and private foundations that sponsor controlled experiments on the other. These include, for instance, the Bureau of Justice Statistics which is responsible for the National Crime Victimization Surveys, and its sister agency, the National Institute of Justice, which has sponsored multi-site controlled experiments on the police handling of domestic violence, among other studies. They include the Bureau of Labor Statistics, an agency that runs large scale surveys on employment and training on the one hand, and the Department of Labor's unit for large scale experiments on residential Job Corps and the Job Training Partnership Act, on the other hand.

The NCES' role as statistical agency, one that is prohibited from doing certain kinds of analytic studies, is complemented by the roles of the Planning and Evaluation Service and OERI at the Office of the Undersecretary at USDE. The PES is required to do evaluative studies that inform policy; OERI's mission is research. Both, at times, use experimental designs that NCES cannot use.

The gap between the statistical agencies and the agencies that focus on analysis represents a kind of intellectual schizophrenia in this country. Data from the former *are*, after all, often used to estimate program effects, not just to describe the condition of education. The insulation of statistical agencies from policy analysis units has considerable political data should be, and, under current laws is, relatively free of political influences. Analysis units that inform policy are more vulnerable to the latter because they must be sensitive to politics. Nonetheless, some have a fine reputation for both independence and political sensitivity. The United States arguably needs to keep the two institutional functions separate. But this does not vitiate the notion that, as an intellectual matter, the separation is unnecessary.

The separation of the statistical agencies and those responsible for analytic studies of programs in education was discussed implicitly and explicitly in a National Academy of Sciences' volume on integrating statistics on children. Brooks-Gunn and colleagues, (1995) and Hoffreth (1995), for instance, recognized the distinctive role of the JOBS experiments and the Perry Pre-School Project in the context of surveys sponsored by NCES and other federal agencies. But they did not explore this deeply. Pallas (1995, p. 153) recognized the merits of NCES and other statistical systems and the distinctive role of experiments on drop-out prevention programs, and more importantly, expressed discomfort with the volume's heavy emphasis on statistical systems. It is a discomfort that we share. It was discussed briefly in a paper on the future of experiments (Boruch, 1994), and is explored here more deeply.

Within some government institutions, the idea of combining different sources of data to estimate relative productivity of programs is not original. For instance, the U.S. Department of Education's Planning and Evaluation Service (PES) has done so often and under a strategic plan (Ginsburg et al., 1992). More to the point, a PES-sponsored evaluation of the Even Start Program for promoting family literacy succeeded in designing and executing studies that are relevant. St. Pierre and colleague (1996) reported on results from a new national information system and from controlled randomized experiments in sites that were embedded in the larger system. This precedent suggests the feasibility of exploiting a design oriented approach to cross-design synthesis within an agency, apart from the approach's exploitation in a cross agency setting.

A FIRST ILLUSTRATION OF THE APPROACH: ADULT LITERACY AND SECOND LANGUAGES

The NCES has undertaken a national probability sample survey of adult literacy in the United States with augmentation for special subpopulations, such as prisoners (Kirsch et al., 1993; Haigler et al., 1994). Suppose that the NCES will run another such survey in the future and that the survey's plan can be influenced.

The U.S. Department of Education's Planning and Evaluation Service has had responsibility for evaluating the effectiveness of certain adult literacy programs. Suppose that another evaluation at multiple sites will be undertaken by this Office and that such plans can be influenced.

Regard the NCES survey on adult literacy and other information obtained by NCES from administrative sources as a data base. Regard the USDE/PES evaluation as a source of controlled experiments. Consider then the question: How can the cross-design synthesis approach inform each agency about the design of new surveys and experiments in the adult literacy arena so as to generate better estimates of the effect of literacy programs in five years?

The cross-design synthesis approach suggests that, in the survey and in controlled experiments, we attend to the:

- target population and samples,
- treatments,
- outcomes, and
- propensity scores.

Each is considered in what follows.

Target Population and Samples

Cross-design synthesis requires that the individuals who are targeted in controlled field experiments are also represented in the survey sample or data base. A new NCES sample survey on adult literacy or English language education in the United States must then include individuals who are targeted for literacy services. Attempts to estimate the effect of the services, undertaken in local controlled experiments, must target similar individuals.

For instance, if one expects that the United States will continue to receive adult immigrants and to focus on adult literacy, the agencies that sponsor literacy programs will continue to focus on such groups. NCES must then plan to periodically direct sample surveys to this same inchoate target population. An evaluation agency, such as PES, ought then to target the same populations in designing experiments that test new approaches to enhancing literacy at a lower cost.

Needless to say, assuring that each attends to the same population will require industry. Perhaps creativity. One must know about unevenness in geographic distributions of migrant groups, such as the Vietnamese in San Diego, Hmong in Minneapolis, and Russians in Brooklyn. And one must take into account the durability or temporariness of such movements.

Treatments

To combine data from experiments and from surveys in cross-design synthesis approach, it is obvious that one must know what treatments (programs) are delivered to whom and when. The implication for design of studies in a new sample survey of literacy in the adult population is that one would have to ask individuals about the literacy or English language programs in which they have participated.

Programs that are all sponsored heavily by a single federal program in literacy might, at the local level, be named as "The Ready Project," "Learn and Earn," or "Newark Knows How to Read." It may be called 50 different names in as many catalogs for junior colleges, evening high schools, or adult education in other venues.

Learning *how* to ask such a question about "treatment" so as to secure reliable responses is difficult, to be sure. Figuring out how to exploit local data bases of literacy services that maintain such information is also likely to be difficult. Nonetheless, NCES must do so if the object is to produce a cross-design synthesis in five years, of who gets what literacy program and to what effect.

For a federal agency or private foundation that sponsors controlled experiments on how to improve literacy programs, the obvious implication is that the agency needs information about the individual's program participation. More important, the method of recording must correspond with how the NCES national survey asks about program participation. Questions about program participation are framed in a survey and the way they are framed in local experiments must be compatible with one another.

Outcomes

The impact of adult literacy programs can be registered partly by measuring an outcome variable such as "literacy level" of each individual or of groups of individuals. To accomplish a cross-design synthesis of the effects of literacy programs in five years, a survey agency such as NCES, must cooperate with an evaluation agency, such as USDE/PES or a private foundation that sponsors evaluations, in developing outcome measures. That is, these organizations must agree on how literacy level is to be measured.

Cooperation of this sort is not easy across local literacy programs, much less across federal agencies or private foundations. For instance, a recurring problem is that local literacy programs, regardless of their sponsorship, have not been able to agree on how to measure literacy. In the absence of agreement, no surveys or experiments undertaken by the federal government are likely to lead to a persuasive cross-design synthesis of whether and which programs work and for whom.

Propensity Scores

A controlled randomized experiment relies on randomization to produce an unbiased estimate of the difference among two or more treatment groups. In such an experiment, individuals who are eligible to be served by a literacy program and who are willing to avail themselves of the program are randomly assigned to the program or to one of two or more variations of the program. Or, entire organizations or communities might be allocated randomly to alternative service programs.

In ordinary language, the treatment groups are "equivalent," apart from chance level difference, because they were randomly composed. A comparison of the groups' performance is then fair. The difference in literacy levels of the two groups following their engagement in the programs or difference in rates of achievement then provide a good estimate of the relative effectiveness of the program variations. Chance variation is taken into account by a sturdy statistical technology.

The NCES does not sponsor studies of the effectiveness of literacy policies or programs. NCES does, however, provide survey data for estimating effects. Statistical analysts who rely on such a platform have usually developed strategies to approximate the results of a controlled experiment, that is, compensate for the absence of the randomized controlled experiment. The strategies vary. During the 1960s, for example, analysts employed OLS estimates of a program effect that was based on a simple, single stage linear model and observational data (e.g., covariance adjustment). Coleman, Hoffer, and Kilgore (1982), exploited the method to construct what amounts to an education production function in understanding the relative effectiveness of private and public schools.

In the 1980s and early 1990s, analysts who depended on path models and more general structural models were able to do persuasive analyses when the underlying theory was well articulated. That is, a causal chain, made explicit through the relevant structural models, could take into account variables in the right sequence and recognize imperfections in measurement. In the 1990s, analysts who employ mixed models and hierarchical linear models (HLM) have been in the ascendancy for good reason. The need to take into account varying levels of institutional influence is obvious.

Each of these approaches tried to take into account the way people find their way into programs, and are selected by others into programs. Each, in some sense, approximates the partial production function one might construct from a massive controlled experiment to determine what influences education productivity and how much.

The focus here is on propensity scores as a device to produce analyses that approximate the results of a controlled test. Such scores were used, apparently to good effect, in the GAO (1995 and Appendix I) report on the relative effectiveness of two approaches to the treatment of breast cancer. The recent work on propensity scores has had the benefit of conscientious thinking about how to recognize the fact that people, in ordinary circumstances, do not engage in programs randomly, and how to incorporate this and related selection factors into analysis.

The GAO's cross-design synthesis of data on treating of breast cancer suggested that the following were important in developing propensity scores in that instance:

- year at which the individual is engaged in treatment,
- geographic area of residence,
- severity of the problem at baseline,
- age of the individual,

- race or ethnicity, and
- marital status.

How and why the variables were chosen was not made plain in the GAO's (1995) report. No theory was presented however. No logic model was articulated.

Similar variables *seem* relevant to understanding production functions. Consider, for instance, the propensity of individuals to engage in adult literacy programs. The access to such programs was greater in the 1990s than it was in 1980. The efforts to entrain clients has arguably been more vigorous in the past few years. Year of engagement then is arguably important. The geographic area of residence and ethnicity are related and theorists argue that it is important to recognize each. For example, Hmong immigrants have clustered in only a few cities, of the west, midwest, and northeast United States. Bosnian and others from the new independent states of eastern Europe make their homes elsewhere.

Marital status may have no obvious influence on a person's inclination to become literate. But a conscientious theorist might argue that adults in the family behave in their children's and family's interest when marriage has occurred. Therefore, the variable called 'marital status' may be a reasonable one to use in constructing a propensity score.

Propensity Scores, Intentions, and Reasons

Roughly speaking, a propensity score reflects the statistical predilection of individuals to (or entities) join or be entrained in one group rather than another. The predilection is indicated by some observable characteristics of the individual. More specifically, it is the conditional probability of being in a particular group given a vector of observed covariates (Rosenbaum & Rubin, 1983, 1984).

For example, high school drop-outs and high school stayers constitute two groups. The probability of being in one group or the other can be characterized descriptively as a function of variables such as daily school attendance rates, age, academic grades, and plans for higher education. Similarly, the probability of entry to college or the work force can be characterized as a function of demographic and other variables.

The variables typically used to estimate a propensity score usually include demographic and contextual information. Rosenbaum (1986), for example, used over 30 such variables to estimate a kind of propensity score for school drop-outs and stayers. They included some of those identified in the paragraph above.

The variables used to compute a propensity score are often "indirect" in the sense that they indicate an individual's state, rather than capturing directly: (a) an intention to belong to one group or another, or (b) the observable reasons for belonging to one group or another. Education surveys, with a few exceptions, do not ask individuals why they dropped out of school or about their intentions to do so.

An implication of the recent analytic work on propensity scores and related analytic methods is that productivity researchers might obtain information on the
individual's intention or on the reasons for membership in a program group. One
rationale for obtaining such information is that it *appears* to be a more direct measure of how people wind up in certain programs than less direct measures, such as
demographic characteristics. The connectedness between an individual's declar-
ing that he/she will drop out of school and the individual's actually dropping out
appears more direct, less distant, from actual membership in the drop-out group
(i.e., becoming a drop-out) than, say, the connectedness between "age" in school
at one point in time and becoming a drop-out in another. Thus far, no formal educational theory has been produced to undergird the construction of propensity
scores. Rather, the justification for the scores lies in small and large sample statistical theory (Rosenbaum & Rubin, 1983).

To the extent that the propensity approach can be informed by education theory
and can help build the theory in a cyclic way, this seems desirable. Better theory,
for example, may promote invention of propensity scores that are easier to interpret. They may decrease the need for a large reservoir of cases on which to match
when propensity scores are used with matching.

Sensible readers can quarrel with the idea that information about reasons or
intentions ought to be elicited in research on education productivity. They do so
with considerable justification. Asking individuals about intentions and reasons is
difficult and, in any case, may not be useful. For instance, Rosenbaum's (1986)
exploration of a propensity score-like approach in a drop-out study using NCES'
High School and Beyond data uncovered the fact that "the vast majority of students who eventually dropped out said in their sophomore year that they expected
to graduate" (p. 208). Was the question asked well? We do not know. We do know
that other "intentions" questions, about aspirations beyond high school, were
indeed used by Rosenbaum to construct the propensity scores.

Some scholars would argue, based on good evidence, that the more general
problem is of understanding revealed preferences and their usefulness in studies
based on observational data. Manski's (1995) book has a nice chapter dedicated to
related matters. The NCES' National Longitudinal Study of the High School Class
of 1972 (NLS-72) served as a vehicle for his attempts to understand how college
enrollment rates would be affected by Pell Grants to needful students. The variables he used as a surrogate for revealed preference included ability, income, and
so on as measured in NLS-72.

Recognizing the skepticism that economists have about self-reported preferences, Manski argued for trying to measure the preferences directly. Part of the
argument is tied to theory, notably theory about what variables to use in an analysis. Economists vary, for example, in the variables they have included in studies of
returns to schooling (p. 97). Some exclude expressed preferences; others include
them. Manski's argument is based also on empirical grounds. He cites research on
consumer buying intentions, fertility (based on Current Population Survey over

the last 50 years), voting intentions, and work by social psychologists to justify his contention that preferences ought to be assessed more directly.

For Manski, one of the implications of agreeing that information on preferences is important in that we must get beyond simple yes-no answers, for example, "Do you think you will drop out of school?" He argues, on analytic and empirical grounds, for eliciting a probabilistic response from each individual. One might then ask: "Looking ahead, what is the probability that you will drop out?" Social psychologists working in the arena would probably go further to argue for eliciting preferences (self-predictions) at points in time that are close to the event in question. Asking in September about students' perceived probability of dropping out is arguably less useful than asking the question in November or December.

Similarly, one may argue that to do a better job constructing propensity scores, one ought to observe or elicit information on why or how people find their way into groups. To return to the adult literacy illustration, NCES might then ask a question of the following sort: "Which of the following factors influenced your decision to enroll (or not enroll) in the literacy program?"

The responses to the question might then be incorporated into a propensity score that is better than (say) one that relies solely on demographic information. Further, the responses may help to develop a small part of a substantive education theory that helps to understand processes by which people enter programs or, more generally, a substantive theory that complements or augments statistical theory for analysis of observational data.

A question of the sort proposed above appears not to have been asked in any large scale observational surveys. Nor can we find concrete illustrations in the published reports on selection modeling, or propensity scores (e.g., Rosenbaum & Rubin, 1983; Rosenbaum & Rubin, 1984; Rosenbaum, 1989). Ways to frame such a question can be developed, based perhaps on NCES expertise and cognitive research in a laboratory or field setting.

To summarize, propensity score approaches suggest that: (a) we consider more seriously whether to measure preference (self-declared propensity), and (b) how and when the preferences are measured seems important. But (c) we need to do research on this and on how to construct propensity scores more generally.

Measurement Issues: Macro-level and Micro-level Studies

In national probability sample surveys, we can often measure a variable using only one or two questions or using an inventory with very few items. Learning about children's relations with other children in a survey might, for example, involve only one or two questions about (say) how many friends that the child says he or she has. A local experiment designed to test ways to improve the ability of withdrawn or hostile children to relate to other children usually involves a more elaborate inventory. It is not clear how to link the data from sparse measures made in a large sample survey to the deeper measures made in the small sample experiment.

Learning about literacy level of individuals in a large sample survey involves a similar problem. One would like to exploit only a few items in a questionnaire so as to reduce the respondent burden. Local experiments, however, can often depend on inventories that demand more time of the program's participants and control group members.

The problem here has a delicious analogue in atmospheric weather research. Satellite imaging might be based on measures on square grids that are 1000 kilometers on edge. Local surface measures may be obtained in far smaller grids. A weather station may yield good data on a grid of 100 kilometers across, for example. This is a more precise local measurement. The challenge lies partly in how to integrate these data across levels of precision of measurement (Draper et al., 1992). Recall also Cronbach's paper on band width versus fidelity.

Learning how to measure simply in large sample surveys and how to measure roughly the same construct with more precision in local experiments are both important. Cross-design synthesis, and the problem of combining different sources of information generally, invites us to learn how to design ways to link the sources in advance of any study.

A SECOND ILLUSTRATION: ABILITY GROUPING IN MATHEMATICS

The following two quotes make plain that the past quarter century has not provided a resolution to debate over the impact of ability grouping on achievement.

> Years of research on [tracking and ability or skill grouping] have not reached stable conclusions. If a student gets into a slow track, he will not be able to catch up with people in faster groups because the pace of the track is lower. On the other hand, maybe he couldn't have kept up anyway...The evidence on tracking goes both ways, and...the intellectual and ideological arguments do the same (Mosteller & Moynihan, 1972, p. 54).

> Appropriate research on skill grouping has not yet been carried out, even though the issues have been debated as public concerns for most of a century (Mosteller, Light, & Sachs, 1996, p. 37).

Research during this period, however, has yielded a better understanding of the prevalence of grouping in a number of academic subjects at several grade levels. The National Assessment of Educational Progress (NAEP and TSA) data indicate, for example, that the incidence of ability grouping increases in most academic subjects as students advance through school. In mathematics, 25 percent of fourth-graders are grouped and 60 percent of eighth-graders are grouped, according to data from 1990 and 1992. (In reading, about 25 percent of fourth-grade students are grouped; this increases to about 33 percent in the eighth-grade.)

Several explanations of why the frequency of ability group increases in the high grades seem plausible. Students of mixed ability, for example, may receive academic instruction in several subjects from one teacher in elementary school (i.e., K-5), reducing the possibility of homogeneous grouping. As students reach the middle school (i.e., 6-8), however, they may receive instruction in several subjects from several teachers. It may then become convenient to group by ability, that is, to arrange heterogeneous groups of students into homogeneous ones. Or, one may posit that differences in achievement accumulate as students age, differences becoming more pronounced in later grades. This may then engender a perceived need for homogeneous grouping. Or, schools may intentionally or otherwise sort students by socioeconomic status, gender, and race, as some critics of ability grouping have charged. Or, decision makers may believe (perhaps on the basis of research evidence) that comparable students, particularly older ones, learn better in homogeneous classes. Hereafter, we attend primarily to the latter two propositions.

This array of processes of grouping has a strong implication for refining surveys such as NAEP so that data can be employed in cross-design synthesis. Recall that one must be able to model the propensity of a person or entity to enter the particular "treatment." In this example, the various rationales that classroom teachers or schools use to justify a grouping strategy might inform the theoretical models for propensity scores.

The array also carries an implication for the design of field experiments whose results would be employed in cross-design synthesis. It is that entire entities that group or do not group students into ability levels might be used as the units of random allocation and analysis. No empirical examples exist. But there are numerous examples of entire classrooms and schools being the unit of randomization in experiments. They have been designed, for instance, to estimate the effect of programs for reducing the use of substances and reducing health risks and understanding the effects of class size. See Boruch (1997) for a brief listing and Boruch and Foley (1998) for a lengthier bibliography, and the examples in the next section.

THE CONCERNS OF EQUAL OPPORTUNITY ADVOCATES

Advocates for equal opportunity often assert that two tracks—one leading to prosperity and the other to poverty—exist in America's schools. That these tracks appear to reflect gender, racial, and socioeconomic differences is cause for alarm. "As a result of the two track system," Beatrix Hamburg, president of the William T. Grant Foundation, writes, "there is educational neglect and underachievement that disproportionately afflicts girls, minorities, and the poor" (1993, p. 9). And "what purpose has desegregation served," Jay Heubert, an attorney and education professor at Harvard University, adds, "if resegregation takes place within desegregated schools?" (personal communication, November, 1992). Ability grouping, from their perspective, may be viewed as one vehicle through which differences along gender, racial, and socioeconomic lines are bred and perpetuated. High quality evidence on whether heterogeneous grouping enhances performance would doubtless illuminate the debate.

RESEARCH EVIDENCE ON ABILITY
GROUPING AND ACHIEVEMENT

Research that produced evidence on a plausible causal relationship between student self-perception and student achievement is often cited as one reason not to group students by homogeneous ability (e.g., see Rosenthal, 1969). The research implies that students in lower ability groups achieve at low levels because they feel inferior or inadequate (relative to their higher grouped peers), and are perhaps treated as such by their teachers.

Oakes (1986a, 1986b) expressed additional concerns about the practice of homogeneous grouping, observing that:

- schools tend to disproportionately place African-American and Latino students in lower ability groups.
- ability groups tend to reflect socioeconomic status, and
- teachers of low-ability groups tend to expose students to fewer, less demanding, topics.

Research on student self-perception and group composition, no matter how interesting on intellectual grounds, does not adequately inform policymakers who are concerned with education's bottom line: "Will heterogeneous grouping raise achievement? And if so, by how much?" Nor does it answer a core question of the discipline of evaluation; namely, "What works? And what works better? For whom?" (Boruch, 1997).

Kulik (1992), Slavin (1993), and Mosteller, Light, and Sachs (1996) have thoughtfully and independently reviewed large bodies of research on grouping in various subjects at different grade levels. Their conclusions differ somewhat—Kulik found grouping to be moderately effective. Slavin found no difference in achievement between homogeneous and heterogeneous grouping. And Mosteller, Light, and Sachs did not offer a firm conclusion on ability grouping's overall impact because they regarded the data as insufficient.

Exploring the matter is complicated partly because of the fragmentary character of data on which we must rely. Consider, for instance, the outline of evidence examined by Slavin (1993) and illustrated in Figure 4.1. The chart shows estimated effect sizes from studies of various kinds on ability grouping.

First, consider that estimating the effect of grouping invites attention to conducting full blown controlled randomized experiments. In question form: If a sample of students were randomly assigned to homogeneous and heterogeneous instructional groups, which group would achieve at a higher level? Asking the question is easy. Mounting randomized experiments has turned out to be more difficult—there have been none in mathematics since 1974, and only two on record. But there have been several non-randomized (i.e., "matched" and "correlational," in Slavin's, 1993 terms) efforts to estimate the impact of homogeneous grouping on math achievement. Slavin (1993) included 16 such studies, plus the two randomized experiments, in his "best evidence synthesis." Slavin found the mean effects of homogeneous grouping to be near zero for the 18 studies.

That more correlational and quasi-experimental studies than randomized experiments have been mounted does not mean the latter are infeasible. It means that they require more resources. More important, it is clear that randomized experiments using entities, such as schools, as the unit of randomization on no less difficult topics has increased over the last decade (Slavin's examples are older). For instance, Boruch and Foley (1998) describe over 30 such projects. There, one finds that in Ellickson and Bell's (1990) multi-state study, about 30 schools were randomized to assay effects of school-wide substance-use prevention programs. No recent experiment involving entities as randomization units focuses on the effect of grouping. But the prospect is tantalizing for any research strategy that seeks to combine results of such experiments with those of observational studies or quasi-experiments. Is there a need to mount such research? Answers may vary. On the one hand, for instance, "...society may not wish to pay for more evidence on this point, but may prefer to decide on its own value system to what extent it regards tracking as acceptable" (Mosteller & Moynihan, 1972, p. 54). But on the other hand, "...the payoff from buckling down to implement a well-designed field study can be high" (Mosteller, Lyte, & Sachs, 1996, p. 6).

A second implication of Figure 4.1 concerns how we think about generalizing the results of an experiment or survey, and the way we relate this task to combining studies. Some researchers (e.g., Elmore, 1993) have questioned the potential of evidence from a "best evidence synthesis" to inform practice. Implicit in the questioning is the notion that a "best-evidence synthesis" (or meta-analysis) contains insufficient evidence to generalize to other settings. This notion is not entirely accurate. Slavin, for instance, includes one analysis (Hoffer, 1992) that made use of data from the Longitudinal Study of American Youth (LSAY), a four-year, large-scale study. He also says that other such studies "provide important additional information not obtainable from the typically smaller and shorter experimental studies" (Slavin, 1993, p. 539). However, Slavin does not discuss the premise underlying cross-design synthesis, namely, that evidence from experimental studies and observational studies might be combined to generate more national estimates of effect. Here, we stress that both survey and designs, and the experiments' designs, must be fixed so as to permit combining results. LSAY, however, may not be the ideal study for this purpose insofar as ability grouping is concerned. Adequate data were available from only 1,800 eighth-grade math stu-

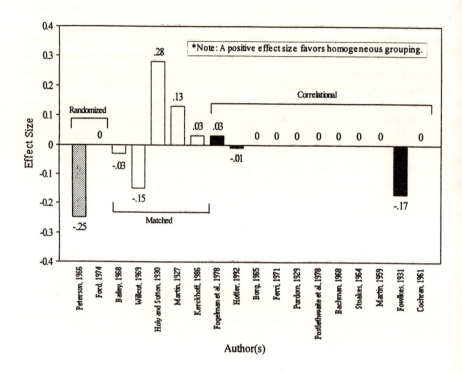

Source: Slavin (1993).

Figure 4.1.   Effect Size Estimates of Middle School Math Studies
that Compared Homogeneous and Heterogeneous Grouping

dents. NAEP's Trial State Assessment, in comparison, collected data from about
2,500 eighth-grade students from *each* state.

WHAT DOES NAEP SEEM TO REVEAL
ABOUT ABILITY GROUPING?

NAEP provides information on student achievement to local, state, and federal
policymakers on a biennial basis. It also provides background information on

students, teachers, and school administrators. Some NAEP information is demographic while other information concerns educational practices and policies. NAEP allows policy makers with data, for example, on whether student achievement is related to ability grouping. NAEP is a passive sample survey, however. Making statements about effectiveness on the basis of NAEP data is therefore inappropriate without some adjustment. It is imprudent to assume, for example, that students who are grouped by ability are comparable in all ways to those who are not. Schools, for example, may tend to group higher achieving students by homogeneous ability rather than heterogeneous ability, thereby causing an imbalance between groups that may bias achievement-based com-

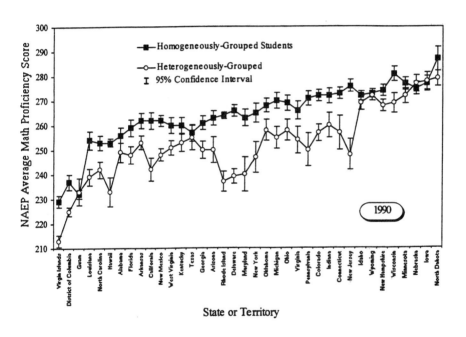

Source: U.S. Department of Education, National Center for Education Statistics (1993, p. 463).

Figure 4.2. Estimates of Math Proficiency for Grade 8 Students by Type of Instructional Grouping (i.e., homogeneous or heterogeneous) for Each State or Territory in 1990

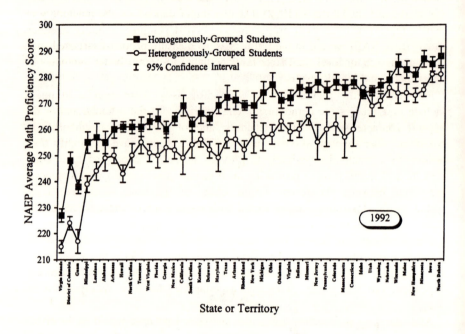

Source: U.S. Department of Education, National Center for Education Statistics (1993, p. 463).

Figure 4.3.   Estimates of Math Proficiency for Grade 8
Students by Type of Instructional Grouping (i.e., homogeneous
or heterogeneous) for Each State or Territory in 1992

parisons. *Unadjusted* NAEP data indicate, to illustrate, that homogeneously
grouped eighth-grade (public school) math students outperformed their hetero-
geneously grouped counterparts in 34 of 37 jurisdictions (significantly in 27 of
37) in 1990, in 43 of 44 jurisdictions (significantly in 34 of 44) in 1992, as Figures 4.2 and 4.3 show.

If one is to use NAEP data to estimate the effects of ability grouping, then one
must first employ a substitute for the randomization of controlled experiments,
that is, to assure that the groups do not differ systematically. The focus here is on
a "propensity score" adjustment—a technique to produce analyses that approximate the results of a controlled experiment. The propensity of schools to choose to

group students by ability, or to not group them, has to be statistically modeled. The model might be based on theory, including local theory. It might be based on earlier work, such as Hoffer's (1992). Or, one might depend on the machinery of logistic regression to estimate propensities without depending explicitly on either. In any case, NAEP would in the future have to measure propensity-related variables at the school level (and state level perhaps) to permit better syntheses of experiments and NAEP results.

For instance, the ingredients for such propensity score estimation might include measures of the demographic characteristics of the schools, and of the general characteristics of their teachers, and of the school neighborhoods. NAEP or other relevant surveys such as TIMSS would have to modify their designs or augment them slightly to provide the data. Deeper and more complex models might involve producing such scores in settings in which students are sorted between classrooms within school on the basis of ability, or are sorted within classrooms, or are sorted both ways. Each of these forces one to attend to differing units of analysis.

This approach can benefit from recent advances in the statistical theory for estimating multilevel models. Bryk and Raudenbush's (1992) hierarchical linear modeling software, for example, enables analysts to model categorical dependent variables while taking into account the multilevel nature of NCES data.

After deriving propensity scores, the analyst's next task would be to divide the entire sample into quintiles on the basis of these scores; that is, to subclassify schools on the basis of their propensity scores. The analyst could then compare the achievement levels of subclassed ability-grouped (homogeneous) and non-ability-grouped (heterogeneous) schools. In a sense, this procedure would generate five estimates of the effect of homogeneous grouping. An example of one possible interpretation is as follows: *With respect to schools that were most likely to group by ability, no difference in achievement exists between those who actually grouped by ability (homogeneously) and those who do not.* The analyst could then combine the estimates by taking the average of the five effects, as in meta-analysis. This final estimate would be far more trustworthy than any of those that were displayed in Figures 4.2 and 4.3.

COMBINING CONTRADICTORY EVIDENCE: A POTENTIAL PROBLEM

How one might combine estimates of effect from experiments and one or more NAEP analyses, particularly when the estimates are contradictory, is unclear. Although the GAO's (1992) introduction to cross design synthesis discusses the problem and presents several options, it concludes that "many refinements are still to be developed" (GAO, 1992, p. 96). The lone illustration (GAO 1995) of a cross design synthesis, however, does not attempt to develop these refinements. The meta-analytic literature, which received a good deal of attention in the GAO's (1992) introduction, also merits consideration here.

In meta-analysis (e.g., see Hedges & Olkin, 1985; Hunter & Schmidt, 1990), the analyst computes one or more overall estimates of effect after (1) collecting, (2) coding, then (3) weighting each study's effect size by its sample size; that is, the analyst computes a weighted average. This weighting scheme poses an analytic problem in the ability grouping example, however, on account of the size and nature of the sample of studies available for analysis. Put into question form: Does each adjusted NAEP state sample (of about 2,500 students) deserve to be weighted by 30 or so more times more than the smallest (Ford, 1974, $n = 82$) experimental study? Probably not.

One approach to the problem is to divide the entire set of studies by design category (i.e., observational, randomized) prior to weighting studies within each category. After doing so, it then seems sensible to follow Hedges and Olkin's advice. The general strategy that they recommend, as applied here, would be to do separate tests of homogeneity for the two sets ([1] 44 state-level NAEP analyses for 1992 and [2] two randomized experiments) of effect sizes. If, say for the NAEP analyses, the null hypothesis of no difference is rejected (i.e., if significant random variation exists among the 44 effect size estimates), the analyst might then include additional covariates in the model to attempt to explain the variation. States with low teacher-student ratios, for example, may produce a small positive effect for ability grouping while those with high ratios may produce a large, negative effect. Teacher-student ratio (e.g., low or high) may therefore account for the variation beyond that expected from sampling error alone among all state-level effect sizes. The analysis, then, would generate two indices of effect for the NAEP analyses. Combining the two, however, would be inappropriate. The analysis would also produce (at least hypothetically because there are only two randomized studies) one or more estimates of effect for the experimental studies. This estimate or these estimates will be distinct from that, or those, produced by the experimental studies.

In this framework, it would also be possible to include in the analysis additional observational studies. Combining Hoffer's (1992) findings on the comparative effectiveness of homogeneous and heterogeneous grouping with those from the potential NAEP analyses, for instance, is one possibility. A re-analysis of Hoffer's data may be in order, however. Although Hoffer makes use of propensity scores, he does not use them to directly compare homogeneous and heterogeneous groups.

This leads to an unpleasantly real implication for both study design and analysis. It is that *it may not be possible* to combine estimates of effect from experiments and one or more state-level NAEP analyses when the estimates contradict one another, particularly when there are very few experimental studies available for synthesis. The risks of combining erroneously are substantial. In medicine, for example, an analysis of 18 meta-analyses and a dozen subsequent clinical trials suggested that relying on the meta-analyses would have led to an ineffective treatment's being adopted in nearly a third of the cases and to effective treatments being rejected at the same rate (Horton, 1997). Exploiting cross-design synthesis in design of studies, can, we believe, avoid such embarrassments and, on the contrary, lead to prouder achievements in understanding.

SUMMARY

Contemporary research on productivity of education institutions has relied heavily on well-designed surveys. Less frequently, it relies on controlled randomized field experiments. The argument and evidence presented here is that experiments and surveys should be designed so as to exploit simultaneously the experiment's chief benefit, notably unbiased estimates of the effects of programs or practices, and the survey's virtue, notably estimates that are more precise in the sense of confidence intervals and more generalizable to large populations than experiments usually are.

The intellectual vehicle for doing so is an approach to data analysis: cross-design synthesis. A special case of meta-analysis, this is a way of arranging how we combine different data sets to make inferences from analyses of data from each source. In this paper, the idea of cross-design synthesis is inverted. That is, we explain how the approach might be employed to *design* experiments and surveys so as to facilitate the eventual joint analysis of their results.

A major assumption underlying this strategy is that political/institutional barriers to design can be handled. In the United States, China, Germany, and other countries, for example, the entities responsible for designing national surveys on which productivity research often depends are different from and organizationally distant from the entities responsible for mounting controlled experiments for evaluating programs (and factors) that are thought to enhance productivity.

Further, we construe the units of analysis more broadly have than in most education production function research. Using students as the units is common, and with good reason. However, using intact classrooms as the unit, or schools, or entire school districts, or groups of students, often make theoretical and practical sense.

The medical examples described briefly here are real. Data from randomized clinical trials and from passive surveillance systems (surveys) have been combined to good effect in research on breast cancer. The education examples are hypothetical simply because inversion of cross-design synthesis has not been applied in this arena. Nonetheless, the contexts, evidence, and opportunity are realistic.

Literacy surveys run in the United States, for example, invite thinking about what one needs to do so as to enrich productivity research. Obviously, in experiments and a survey, target populations would have to overlap. So one would have to assure that the measures of outcomes are the same and measured well in each, and that what constitutes "treatment" programs is defined the same way and measured in each. Less obviously, one needs to invent and measure variables in the survey that permit construction of propensity scores—the likelihood that an individual will wind up in one literacy program or another or none at all.

In a second example, on analyses of student achievement as a function of ability grouping is based on the National Assessment of Educational Progress. Generalizations to other surveys, such as the next version of the Third International Mathematics and Science Study (TIMSS), or to new longitudinal surveys, are obvious. The NAEP-based illustration emphasizes that surveys such as NAEP obtain data on grouping, but that few controlled experiments have been done. More important perhaps, better work might be done using classrooms or schools as the unit of random allocation in experiments, given that these constitute the first stage sampling units in TIMSS and NAEP respectively. Here too, some hard work is needed to understand how to build propensity scores based on entities rather than individuals as the units of analyses in productivity research.

ACKNOWLEDGMENT

Work on this topic was supported through small contracts by the U.S. Department of Education, notably the National Center for Education Statistics and the USDE Planning and Evaluation Service. We are grateful to each for sponsoring the exploration of ideas. This paper is also based partly on reports produced for the U.S. Department of Education (1996).

NOTE

1.   The definition and illustrations in this paper focus on randomized experiments, partly because this is a scientifically sturdy anchor point. The reader will recognize that some other designs for production function research will also yield unbiased estimates of relative differences or of program effects. But they require a different kind of active control of the process of assigning individuals or entities to a program. Regression-discontinuity designs are a case in point. See Cahan and Linchevevski (1996) for an illustration of the approach in the context of a longitudinal study.

REFERENCES

Bickman, L. (1985). Randomized field experiments in education. *New Directions for Program Evaluation, 28*, 39-54.
Boruch, R. B. (1975). Coupling experiments and approximations to experiments in social program evaluation. *Sociological Methods and Research, 4*, 31-53.
Boruch, R. F. (1994). The future of controlled experiments. *Evaluation Practice, 15*(3), 265-274.
Boruch, R. F., & Terhanian, G. (1996). So what? The implications of new analytic methods for designing NCES surveys. In G. Hoachlander, J. Griffith, J. H. Ralph, & E. McArthur (Eds.), *From*

data to Information: New Directions for the national Center for Education Statistics (Chap. 4). Washington, DC: U.S. Department of Education, National Center for Education Statistics, USDE NCES 96-901.

Boruch, R. F. (1997). *Randomized experiments for planning and evaluation: A Practical Guide.* Thousand Oaks, CA: Sage.

Boruch, R. F., & Foley, E. (1998). *Sites and other entities as the unit of allocation and analysis in controlled experiments.* Research Report No. P-604 Philadelphia, PA: University of Pennsylvania Center for Research and Evaluation in Social Policy.

Brooks-Gunn, J., Brown, B., Duncan, B., & Moore, K. A. (1995). Child development in the context of family and community resources. In *Integrating federal statistics on children: Report on a workshop* (pp. 27-97). Washington, DC: National Academy Press.

Bryk, A. S., & Raudenbush, S. W. (1992). *Hierarchical linear models: Applications and data analysis.* Newbury Park, CA: Sage.

Cahan, S., & Linchevski. (1966). The cumulative effect of ability grouping on mathematical achievement: A longitudinal perspective. *Studies in Educational Evaluation, 22*(1), 29-40.

Charlton, B. G., Taylor, R. P. A., & Procter, S. J. (1997). The PACE strategy: A new approach to multi-centered clinical research. *Quarterly Journal Medicine, 90*, 147-151.

Coleman, J. S., Hoffer, T., & Kilgore, S. (1982). *High school achievement: Public, catholic, and private schools compared.* New York: Basic Books.

Davis, C., & Sonnenberg, B. (Eds.) (1995). *Programs and plans of the National Center for Education Statistics: 1995 edition.* Washington, DC: NCES, U.S. Department of Education.

Draper, D. Graves, D. P., Goel, P. K., Greenhouse, J., Hedges, L. V., Morris, C. N., Tucker, J. R., & Waternaux, C. M. (1992). *Combining information for research.* Washington, DC: National Academy of Sciences. Also in: *Contemporary Statistics*, No. 1. Alexandria, VA: American Statistical Association, Undated.

Droitcour, J. A., & Chelimsky, E. (1995). *Cross-design synthesis.* Paper presented at the Annual Meeting of the American Evaluation Association. Vancouver, Canada.

Droitcour, J. A., Silberman, G., & Chelimsky, E. (1993). Design synthesis. *International Journal of Technology Assessment in Healthcare, 9*(3), 440-449.

Ellickson, P. L., & Bell, R. M. (1990). Drug prevention in junior high: A multi-site longitudinal test. *Science, 247*, 1299-1305.

Elmore, R.F. (1993). What knowledge base? *Review of Educational Research, 63*, 314-318.

Engels, E. A. & Spitz, M. (1997). PACE-setting research. *Lancet, 350*(9079), 677-628.

Fienberg, S. B., & Tanur, J. (1987). Experimental and sampling structures. *International Statistics Review, 55*, 65-96.

Finn, J. D., & Achilles, C. M. (1990). Answers and questions about class size: A statewide experiment. *American Education Research Journal, 27*, 557-577.

Ginsburg, A. et al. (1992). Reinvigorating program evaluation in the U.S. Department of Education. *Educational Researcher, 21*(3), 24-27.

Greenwald, R., Hedges, L. V., & Laine, R. D. (1996). The effect of school resources on student achievement. *Review of Educational Research, 66*(30), 361-396.

Haigler, K. O., Harlow, C., O'Conner, P., & Campbell, A. (1994). *Literacy behind prison walls.* Washington, DC: National Center for Education Statistics.

Hamburg, B. (1993) *New futures for the forgotten half: Realizing unused potential for learning and productivity: William T. Grant foundation annual report.* New York: The William T. Grant Foundation.

Hanushek, E. A. (1989). The impact of differential expenditures on student performance. *Educational Researcher, 18*(4), 45-52.

Hedges, L. V., & Olkin, I. (1985). *Statistical methods for meta-analysis.* Orlando, FL: Academic Press.

Hoffer, T. B. (1992). Middle school ability grouping and student achievement in science and mathematics. *Educational Evaluation and Policy and Analysis, 14*(3), 205-227.

Hoffreth, S. (1995). Children's transition to school. In Board on Children and Families and Committee on National Statistics (Ed.), *Integrating federal statistics on children: Report of a workshop* (pp. 98-121). Washington, DC: National Academy Press.

Horton, F. (1997). Editorial: Meta-analysis under scrutiny. *Lancet, 350*, 675.

Hunter, J. E., & Schmidt, F. L. (1990). *Methods of meta-analysis*. Newbury Park, CA: Sage Publications.

Kulik, J. A. (1992). An analysis of the research on ability grouping: historical and contemporary perspectives. *Ability Grouping Research-Based Decision Making Series, Number 9204*. Ann Arbor: The University of Michigan Press.

Kirsch, I. S., Jungeblut, A., Jenkins, L., & Kolstad. (1993) *Adult literacy in America*. Washington, DC: National Center for Education Statistics USGAO #065-000-00588-3.

Manski, C. R. (1995). *Identification problems in the social sciences*. Cambridge, MA: Harvard University Press.

Mosteller, F. & Moynihan, E. (Eds.) (1972). *On equality of educational opportunity*. New York: Vintage Books.

Mosteller, F., Light, R., & Sachs, J. (1996). *Sustained inquiry in education: Lessons from ability grouping and class size*. Cambridge, MA: Center for Evaluation of the Program on Initiatives for Children, Harvard University.

Myers, D., & Schirm, A. (1997). *The short term impacts of upward bound: An interim report*. Princeton, NJ: Mathematica Policy Research Contract No. LC-92001001.

Oakes, J. (1989a). Keeping track, part I: The policy and practice of curricular inequality. *Phi Delta Kappan, 68*, 12-17.

Oakes, J. (1989b). Keeping track, part II: Curriculum inequality and school reform. *Phi Delta Kappan, 68*, 148-154.

Oakes, J. (1990). *Multiplying inequalities: The effects of race, social class, and tracking on opportunities to learn mathematics and sciences*. Santa Monica, CA: Rand Corporation.

Pallas, A. (1995). Federal data on educational attainment and the transition to work. In Board on Children and Families and Committee on Federal Statistics (Ed.), *Integrating federal statistics on children: Proceedings of a workshop* (pp. 122-155). Washington, DC: National Academy Press.

Procter, S. J., Taylor, P. R. A., Stark, A. et al. (1995). Evaluation of the impact of allogenic transplant in first remission on an unselected population of patients with acute myeloid leukemia, aged 15-55 years. *Leukemia, 9*, 1246-1251.

Reynolds, A. J., & Walberg, H. J. (1993). Structural modeling of nine factors of educational productivity. In H. J. Walberg (Ed.), *Advances in educational productivity* (Vol. 3, pp. 79-95). Greenwich, CT. JAI Press.

Rosenbaum, P. R. (1986). Dropping out of high school in the United States: An observational study. *Journal of Educational Statistics, 11*(3), 207-224.

Rosenbaum, P. R. (1989). Optimal matching for observational studies. *Journal of the American Statistical Association, 84*(408), 104-1032.

Rosenbaum, P. R., & Rubin, D. B. (1983). The central role of the propensity score in observational studies for casual effects. *Biometrika, 70*(1), 41-55.

Rosenbaum, P. R., & Rubin, D. B. (1984). Reducing bias in observational studies using subclassification on the propensity score. *Journal of the American Statistical Association, 79*(387), 516-524.

Slavin, R. E. (1993). Ability grouping in the middle grades: Achievement effects and alternatives. *Elementary School Journal, 93*(5): 535-552.

St. Pierre, R. G., Swartz, J., Gamse, B, Murray, S., & Deck, D., (1996). *Improving family literacy: Findings from the national even start evaluation*. Cambridge, MA: Abt Associates.

Terhanian, G. (1996). Applications of new methods of analysis to data from the national educational longitudinal study of 1988. *NELS*, 88. http://dolphin.upenn.edu/~terhania/

U.S. Department of Education. (1995). *National center for education statistics programs and plans: 1995*. Washington, DC: NCES.

U.S. Department of Education. (1996). *From data to information: New directions for the national center for education statistics*, NCES 96-901, by G. Hoachlander, J. E. Griffith and J. H. Ralph, E. McArthur, project officer. Washington, D.C.

U.S. Department of Education, Planning and Evaluation Service, Office of the Under Secretary (1997). *The short term impacts of upward bound: An interim report*. Washington, DC: USDE.

U.S. General Accounting Office. (1992). *Cross design synthesis: A new strategy for medical effectiveness research*. Washington, DC: USGAO GAO/PEM D-92-18.

U.S. General Accounting Office. (1995). *Breast conservation versus mastectomy: Patient survival in day-to-day medical practice and in randomized studies*. Washington, DC: USGAO GAO/PEM D-95-9.

Walberg, H. J. (1993). Introduction and overview. In H. J. Walberg (Ed.), *Advances in educational productivity* (Vol. 3, pp. 3-20). Greenwich, CT: JAI Press.

Wilson, D. B. (1995). *The role of method in treatment effect estimates: Evidence from psychological, behavioral, and educational meta-analyses*. PhD Dissertation. Claremont Graduate School, Claremont, CA.

Chapter 5

STAKEHOLDER CONCEPTS

Marvin C. Alkin, Carolyn H. Hofstetter, and Xiaoxia Ai

This chapter was commissioned originally on the topic of "THE Stakeholder Approach to Evaluation." There is, in fact, an evaluation approach described within the literature, which was specifically called stakeholder evaluation. In the mid-1970s, the National Institute of Education (NIE), concerned about the shortcomings of educational evaluation documented in the literature (Guba, 1969), designed and implemented a "stakeholder approach" for use in evaluating two large national programs (Gold, 1983). This work received considerable visibility at that time.

Since then, the idea of stakeholder participation has permeated a number of prominent views which have influenced evaluation practice. Their contention is that stakeholder participation dramatically improves the quality of evaluations. Despite the consensus for including stakeholders, evaluators vary widely in the rationales employed for stakeholder participation. Greene adds that "for most of these rationales, neither the theoretical nor the operational elements of stakeholder participation are very well developed or understood" (1986, p. 3).

Advances in Educational Productivity, Volume 7, pages 87-113.
Copyright © 1998 by JAI Press Inc.
All rights of reproduction in any form reserved.
ISBN: 0-7623-0253-4

The origins of stakeholder ideas are not new. Indeed, Weiss (1983a, p. 6) has noted that evaluators had for a number of years been urged to involve users (and user groups) in the evaluations they conduct and many evaluators at least paid lip service to this advice. House places the notion of stakeholders within a broader perspective, noting that evaluation may be seen as a political activity and the consequences of that activity have implications for evaluation:

> In the early days it was assumed that the interests of all parties were properly reflected in the traditional outcome measures, but this assumption came to be questioned, and it was recognized that different groups might have different interests and might be differentially affected by the program and its evaluation. "Stakeholders" (those who had a stake in the program under review) became a common concept, and representing stakeholder views in the evaluation became an accepted practice (House, 1993, p. 9).

Thus, the notion of stakeholder participation received broad credence and has become widely incorporated into a number of approaches to evaluation. This chapter provides an examination of the seemingly simple, yet complex, character of the stakeholder concept in evaluation. A number of evaluation theoretic points of view will be examined in order to demonstrate the way in which stakeholder concepts are incorporated into each.

REASONS FOR STAKEHOLDER INVOLVEMENT

Evaluation theorists justify their advocacy of stakeholder involvement in many ways. The range of these views extends from moral/philosophical (it is the right thing to do) to functional, which cites particular benefits to be derived. Further, regardless of the rationale for involving stakeholders in the evaluation, the theory and operationalization of stakeholder participation may differ for each rationale (Greene, 1986).

Some assert the necessity of stakeholder involvement as an appropriate part of a democratic system. It is maintained that stakeholder participation provides an appropriate recognition of diverse values and interests in society and assures that programs will not be dictated by only the bureaucrats and those in power, but also by lower status stakeholders. This argument is of great concern to theorists like House (1993), Guba and Lincoln (1981), and Fetterman (1996).

At a more functional level, stakeholder involvement is deemed to be justified based upon three particular benefits assumed to derive from stakeholder involvement. The first of these is the improved relevance of the evaluation. It is presumed that involvement of the broad base of stakeholders improves the relevance of an evaluation because all appropriate points of view have been registered. That is, participation and the registering of one's views are thought to impact on the goals, structure, and procedure. Thus, improved relevance is attained.

Second, there is a strong belief among evaluation theorists that stakeholder involvement not only increases the relevance of the evaluation information but also the subsequent commitment to the evaluation. The view is that to the extent to which stakeholders are involved, they place greater trust in the evaluation and are consequently committed to it. This commitment takes several forms. On the one hand, commitment may be expressed by actions during the course of an evaluation that facilitate its conduct. Practicing evaluators can easily document instances in which lack of cooperation on the part of various stakeholder groups greatly impeded evaluators' abilities to conduct their work. Commitment to an evaluation is also expressed related to evaluation findings. Assumptions related to the stakeholder process maintain that involvement and participation of stakeholders increase the likelihood that they will be committed to the evaluation findings (or, should we instead believe that the more you know about something the greater your basis for not trusting it?).

The most widely maintained view about the benefits of stakeholder involvement relates to the perception of enhanced evaluation use. Weiss states simply "wider and more immediate use of evaluation results is the driving force behind the stakeholder model" (1983a, p. 8). This belief about the consequences of stakeholder participation simply completes a logical sequence: stakeholder involvement improves the relevance of evaluation which in turn produces a higher level of commitment on the part of stakeholders leading to increased use of evaluations. Patton summarizes nicely the point of view related to increased use.

> [The literature documents] the proposition that people are more likely to accept and use information, and make changes based on information, when they are personally involved in and have a personal stake in the decision-making processes aimed at bringing about change. Most directly, there is a growing evaluation and policy analysis literature—an empirical literature— that supports the proposition that utilization of evaluation is enhanced by high-quality stakeholder involvement in and commitment to the evaluation process (Patton, 1988, p. 325).

WHO ARE STAKEHOLDERS?

There are a variety of definitions of who are "stakeholders." In part, these definitions emanate from evaluators' particular theoretic point of view. Gold (1983, p. 64) who developed the "stakeholder evaluation" procedure at the National Institute of Education defines stakeholders simply "as individuals with a vested interest in the outcome of evaluations." Weiss provides a more elaborate description:

> I interpret the term *stakeholders* to mean either the members of groups that are palpably affected by the program and who therefore will conceivably be affected by evaluative conclusions about the program or the members of groups that make decisions about the future of the program, such as decisions to continue or discontinue funding or to alter modes of program operation (1983b, p. 84).

As the term stakeholder has become more entrenched within the literature, its appeal has attracted the broad panoply of theorists who previously may have employed other terms. Thus, we note that in more recent editions of textbooks by Rossi and Freeman (1993) and by Patton (1997), for example, the term "stakeholder" is now more prominently mentioned. Patton now notes that the process of "identifying people who can benefit from evaluation is so important that evaluators have adopted a special term for potential evaluation users: *stakeholders*"(1997, p. 41). Thus, Patton's definition of stakeholders is an extension of the group he has long identified as an essential part of his evaluation procedure: namely, "primary intended users." Stakeholders, thus, form a larger group from whom primary intended users will be selected.

Who is this broad audience of stakeholders? Generally, the definition and selection criteria for stakeholders differ based on the guiding rationale(s) for the evaluation (Greene, 1986). Guba and Lincoln (1981) categorize stakeholders into three broadly defined and diverse groups: (1) people involved in developing and using the evaluand (program developers, funders, current, and future adopters); (2) direct and indirect beneficiaries of the evaluand; and (3) people suffering disadvantage related to the evaluand (groups excluded from participation). In contrast, Weiss (1983b) categorizes stakeholders into four more specifically defined groups: policymakers, program managers, practitioners, and clients and citizen organizations. The specifics of which stakeholders are included within each category are defined largely by the nature of the particular evaluation.

While there is widespread agreement about the appropriateness of Weiss' categories of stakeholders, there are nonetheless areas of potential disagreement. For example, Stake (1983) quotes House as saying "the government cannot be a stakeholder." Most writers who have discussed the notion of stakeholders view program administrators at all governmental levels as part of the stakeholder group.

Others would extend the stakeholder group beyond those mentioned above. Patton, for example, maintains that the utilization-focused evaluator is also a stakeholder: "the active-reactive-adaptive process connotes an obligation on the part of the evaluator to represent the standards and principles of the profession as well as his/her own sense of morality and integrity...." (1997, p. 364). House would clearly concur, noting the obligations of evaluators to bring moral concerns about social justice into the process. Sechrest, Babcock and Smith (1993, p. 231), however, caution that "evaluators must be very careful lest their values dominate the evaluation process."

HOW ARE STAKEHOLDERS SELECTED?

As previously noted, how stakeholders are defined and the rationale(s) for an evaluation guide largely the selection criteria for stakeholders. Greene (1986) presents several examples: (1) for a utilization rationale, program knowledge and status as a potential evaluation user may be important selection criteria; (2) for an empowerment rationale, low hierarchical status; and (3) for a decision making rationale, status and power as a decision maker. Further, Greene (1986, p. 3) notes that the type of stakeholder participation varies based on the different rationales. For example, meaningful participation in a utilization rationale may be defined as "stakeholder control over decisions about substance or content."

Value judgments, rather than technical judgments, also play an important role in stakeholder selection and participation. Mark and Shotland (1985, p. 607) note that *"whose questions* will guide an evaluation can be seen as essentially an issue of values...That is, if a particular stakeholder group is allowed to participate effectively, its value preferences will determine the focus of the evaluation."

Other theorists advocate different bases for making choices on stakeholder inclusion or emphasis. Weiss (1983a, p. 9) has noted that, "Having a stake in a program is not the same thing as having a stake in an evaluation of a program." The fact that a great number of stakeholders have been identified as having an interest related to a program does not mean that they are potentially interested in the results of an evaluation. Patton expands on the theme by seeking to identify from among the broader stakeholder group, those whom he terms "primary intended users." His thesis is that primary stakeholders should be those whom have a high stake in the evaluations' outcomes and a strong interest in the results of the evaluation. He has developed a matrix for mapping stakeholders' stakes (see Patton, 1997, p. 344).

Thus, the procedure employed by evaluators for deciding on stakeholder inclusion is framed by the views of each theorist and the rationale for a particular evaluation. The intended purpose of the evaluation (e.g., social justice, participation, utilization) forms the basis for choice decisions. This is understandable but also raises a dilemma. Weiss (1983a, p. 10) frames this dilemma nicely in the form of a question: "...does the right to decide who is in and who is out reduce the efficacy of stakeholder evaluation as an instrument of democratization?"

Recognizing that stakeholders will have different questions and that constraints must be posed in an evaluation, an obvious question for an evaluator is *"which* stakeholders' concerns will be addressed?" Mark and Shotland (1985, p. 609) provide insight into this dilemma by classifying stakeholder groups into two dimensions: the perceived power and the perceived legitimacy of the group's interests. They note that the evaluator generally rates potential stakeholder groups along these two dimensions (high/low) as influenced by: (1) the evaluator's characteristics (e.g., evaluator's theoretical perspective), (2) the evaluator's role relative to various stakeholders (e.g., is one stakeholder paying for the evaluation?); and (3) the purpose of the evaluation (e.g., external accountability, such that an external audience might be most powerful?). For example, Mark and Shotland (1985) note that an empowerment rationale might imply the selection of low-power, high-legitimacy stakeholder groups.

CAN ALL STAKEHOLDERS BE INCLUDED?

Weiss (1983a, p. 9), in commenting on the American Institute of Research's (AIR) evaluation of the Cities in Schools Program (CIS), summarizes the groups identified as stakeholders including "teachers, school administrators, students, parents, school board members, the mayor's Office, the Congress, the research community, the Department of Education, state officials, national program managers, and community organizations." She raises the obvious question of whether a single evaluation study can possibly satisfy such disparate audiences. Specifically, Weiss (1983a, p. 9) asks, "Can any team of evaluators, however skillful, provide information of immediate utility for various sets of program participants and at the same time, conduct an adequate test of a replicable program model?" We would add that the potential utility of satisfying multiple stakeholders applies irrespective of whether or not the evaluator is testing a specific program model.

In examining the same AIR evaluation, Stake (1983, p. 25) noted, "The attempt to be useful to many may in fact have prevented it from being useful to any." We are reminded of the biblical adage "no man can serve two masters." Indeed, when we consider the great multiplicity of stakeholders that some might wish to include in an evaluation, the situation is greatly exacerbated beyond serving only two masters.

How many stakeholders is too many? Cousins, Donohue, and Bloom (1996) in a survey of evaluators found that six stakeholders was reported as the median number typically involved in a project. It is difficult to interpret what this means. Did those surveyed count as different stakeholders even if they were individuals within the same stakeholder group? Alternatively, if stakeholder groups were each counted as a single stakeholder, how does an evaluator determine the views of that group? Various evaluation theorists suggest means of reaching consensus within stakeholder groups and of combining stakeholder views across groups. These will be addressed within a subsequent discussion.

Given the multitude of ways to define stakeholder constituencies, there is nonetheless the necessity for making choices. A number of theorists have attempted to make choices as to how multiple stakeholder perspectives can be combined or weighted. For example, Shadish, Cook, and Leviton (1991), advocate the use of multiple synthesis—in essence: "If x then y" value summaries.

House (1995, p. 4) rejects multiple summaries and takes the position that evaluators "can usually arrive at legitimate single (if highly qualified) judgments about the values and interests of stakeholders in an evaluation." In doing this, he would maintain that the evaluator has a special responsibility for seeking out the views of stakeholders from minority groups and from the poor and powerless. He advocates this special consideration to compensate for the likelihood that individuals from the majority culture are "the most powerful players" and would prevail in the discussion of stakeholder interests (House, 1993, p. 157). His view is that even approaches such as Fourth Generation Evaluation (Guba & Lincoln, 1989) which attempt to incorporate minority views and interests as part of the stakeholder process would disadvantage minority groups in the face-to-face negotiations, which are part of that theoretic stance. In such instances, House suggests that minorities and poor would succumb to the views of the powerful.

NATURE AND EXTENT OF STAKEHOLDER INVOLVEMENT

What does it mean to "involve" a stakeholder? An examination of the evaluation literature demonstrates highly different levels of engagement perceived as stakeholder involvement. On the one hand, the more research oriented approaches to evaluation consider stakeholder involvement as primarily attempting to perceive the needs of the various constituencies and to be responsive to those needs. Alternatively, the more participant oriented approaches to evaluation prescribe an active engagement of stakeholders in a variety of essential components of conducting an evaluation.

Stakeholders can be involved in evaluations in a number of ways. The nature of this potential participation mirrors the various stages of the conduct of an evaluation. Stakeholders may be a source of problem identification—what is to be evaluated. They might participate in the development of the design and methodology. They could also provide input into the data analysis and, more importantly, participate in determining the way to place values on data. Finally, stakeholders might be designated targets of the evaluation reporting. Thus, a discussion of stakeholder involvement in evaluation must examine the extent to which various evaluation theoretic positions consider stakeholder participation with respect to each of these dimensions.

The first stage in the conduct of an evaluation has been referred to as the "evaluation focusing" activity. Alkin (1987) has referred to this as the most important step in the evaluation process. Decisions about the purpose of the evaluation, the reasons for its conduct, and the ways in which it might benefit the entities being evaluated are all part of an evaluation focusing activity. There is great variability in the way in which evaluators engage stakeholders in the process of determining evaluation purposes and intents. Furthermore, there are great variations between evaluation approaches in the diversity of stakeholders whose participation is elicited. Some theorists may define stakeholders quite narrowly or elicit varying levels of participation from different stakeholder groups. Some stakeholders' views may be sought through a questionnaire (for example) while actively engaging a smaller set of stakeholders in the problem identification. Furthermore, there is the issue of the extent to which the evaluation issues and direction are solely established at the initiation of the evaluation or whether they are subject to continuing modification. If the latter, to what extent are stakeholders involved in decisions related to modified evaluation expectations?

Many evaluation theorists, particularly those with more traditional evaluation orientations, view the development of the design and methodology as totally the purview of the evaluator. They argue that the evaluator has methodological expertise which would dictate that this phase of an evaluation is solely the evaluator's responsibility. Another point of view would contend that while the evaluator certainly has a particular expertise, the engagement of stakeholders even in this phase adds insights into the practicality and feasibility of activating the design. Moreover, stakeholder participation in this stage potentially enhances their commitment to the evaluation. Furthermore, as we will note in the subsequent discussion, some evaluation approaches view participation in the development of design and methodology as centrally related to the purpose of evaluation—the improvement of stakeholder skills.

Scriven has historically made a major point of emphasizing the unique (and singular) role of the evaluator in "valuing" data (1993). He and other theorists of a like mind maintain that the valuing of data is strictly within the domain of the professional evaluator. Typically they cite the expertise of the evaluators and the fear that bias might be introduced by allowing stakeholders to participate in this process. Other evaluation points of view consider stakeholder participation in making judgments about data an essential part of the process. They might argue that bias is introduced by allowing the single perspective of the evaluator to be employed in the valuing of data and that, in fact, all data are subject to interpretation based upon multiple realities. Thus, it would follow from this argument that participation of stakeholders in valuing is essential. Some theorists continue this theme by stressing evaluation as an activity intended to empower stakeholders and view the act of valuing as an integral part of that empowerment.

Stakeholders are also the focus of evaluation reporting. All evaluation theorists presume that evaluations will be reported, in some manner, to individuals or audiences. At issue is the nature of the stakeholders to whom evaluations are presented (which stakeholders), the extensiveness of stakeholder reporting (how many stakeholder groups), and the relative importance of various stakeholder groups in the reporting process (primary or secondary).

SOME EVALUATION THEORETIC POSITIONS

As noted, different views on the nature of evaluation have impact on the role of stakeholders. This will be illustrated by describing several evaluation theoretic positions in which stakeholders play a prominent role.

Stakeholder-Based Evaluation

As discussed earlier, Gold and his colleagues at the National Institute of Education (NIE) sometime ago developed a "stakeholder approach" explicitly to empower a variety of prospective users (defined as stakeholders) to participate actively in the evaluation process. It was presumed that the involvement would increase both the evaluation utility and the credibility of the evaluation for decision makers.

Gold claims that concern for evaluation utility was the driving force behind the design/development of this stakeholder model. The aim of increased use was to be achieved by bringing a wider variety of people into active participation in the design and conduct of the evaluation. According to Gold (1983, p. 64), "Stakeholder groups contribute to the evaluation process by specifying the kinds of evaluative information required; and the most useful form for presentation of the information." He maintains that the evaluator is responsible for furnishing the evaluation method most responsive to the user specified evaluation information needs, and for providing periodic feedback so that stakeholder involvement is sustained throughout an evaluation study.

Gold suggests that the determination of an initial set of expectations is done in a deliberately informal way by gathering representative views from a variety of stakeholders (which can include national, local, program, and participant interests). At the first meeting, stakeholders should describe their expectations for the program, state their requirements for evaluation data and when they need this information, and suggest the most useful form in which evaluation can be presented to them. Thus, in this stakeholder-based evaluation, stakeholders are actively involved in problem identification from the beginning.

Stakeholders' continuing involvement in problem identification, according to Gold's descriptions, is sustained by continual re-examination of expectations for program progress. The NIE stakeholder approach asks users to identify their expectations and assumes that these expectations are changeable. The process of determining and testing expectations is both dynamic and interactive. Stakeholders are provided with periodic feedback which includes information assessing program progress. If expectations are not met, stakeholder groups and the evaluator must decide whether to modify them or to alter strategies. Gold insists that this process continues until stakeholders feel comfortable with the way in which the program is operating.

In the NIE model, stakeholders also have input into the development of the evaluation design. Gold claims that stakeholder-based evaluation requires the evaluator to develop a design "that is responsive to stakeholders' needs, that is fair to the program in the sense that expectations are reasonable, and that is workable within the limitations of evaluation technology" (Gold, 1983, p. 66). Stakeholders do not participate in developing the design, although, the design is not implemented unless it is reviewed by stakeholder groups (including the funding agency) and supported by sufficient numbers of stakeholders.

Gold states that stakeholders, as a focus of evaluation reporting, must specify the kinds of evaluative information they want, the form in which evaluative information is to be reported, and when they need them. The reports usually are short briefing summary papers about the state of the program at the time requested by stakeholders, or they often can be verbal reports of major findings related to the state of the program. This, according to Gold, is consistent with the evaluator's obligation "to respond to stakeholder requests with the most recent information available and in the form requested" (Gold, 1983, p. 66). Stakeholders are the audience of reporting but do not actually participate in report preparation or review.

Utilization-Focused Evaluation

Patton in his utilization-focused evaluation approach presents a procedure which also emphasizes stakeholder involvement (1986, 1997). Patton is concerned with maximizing the likelihood that evaluations will be used. He postulates that this outcome can most likely be attained if evaluators focus their efforts on individuals who have real interests in the findings of the evaluation and a likelihood of using the evaluation to improve programs.

In utilization-focused evaluation, the first step is to identify and analyze the needs of various stakeholders of the evaluation. Patton defines stakeholders broadly, noting (1997, p. 42):

> Evaluation stakeholders are people who have a stake—a vested interest—in evaluation findings. For any evaluation, there are multiple possible stakeholders: program funders, staff, administrators, and clients or program participants....Stakeholders include anyone who makes decisions or desires information about a program.

However, as the stakeholders will have multiple and varying degrees of interest in any given evaluation, the evaluator must narrow choices to relevant and specifically identifiable stakeholders, the "real" primary intended users, rather than vague, passive audiences (Patton, 1997, p. 365). Thus, the initial identification of specific decision makers and information users who are genuinely interested in the findings of the evaluation is fundamental. The purpose of the evaluation is to answer the stakeholders' questions, while being sensitive to and respectful of the varied and multiple interests (Patton, 1986).

Selected stakeholders are actively involved in every step of the evaluation process. They are involved, through personal engagement, in the complex processes of goals clarification, issues identification, operationalizing outcomes, matching research design to program design, determining sampling strategies, organizing data collection, interpreting results, and drawing conclusions. Patton notes (1988, p. 326):

> These processes take stakeholders through a gradual awakening to program complexities and realities, an awakening that contains understandings and insights that will find their way into program developments over time, only some of which will be manifested in concrete decision. Utilization begins as soon as stakeholders become actively involved in evaluation because that involvement, properly facilitated, forces them to think about program priorities and realities.

The stakeholder assumption, then includes the expectation that stakeholders need to extend time and effort to figure out what is worth doing in an evaluation, they need help in focusing on worthwhile questions; and they need to experience the full evaluation process if that process, which is really a learning process, is to realize its potential, multi-layered effects.

In identifying the problems for study, the utilization-focused evaluator may find out what various stakeholders think about and how they define evaluation. Then, the evaluator may ask a series of questions designed to focus the evaluation, identify the questions and stakeholders at hand, and to determine which primary intended users and issues will serve to focus the evaluation.

This level of high quality participation is also evident in stakeholder involvement in methodological decisions. It is the role of the evaluator to inform and teach stakeholders about various methodological issues, in an understandable manner, so that they may choose among the strengths and weaknesses of different design and methodological options. Patton notes (1988, p. 315) that "such involvement in collaborative deliberations on methodological issues can significantly increase stakeholders' understanding of the evaluation, while giving evaluators a better understanding of stakeholder priorities and situational constraints on the feasibility of alternative approaches."

Patton also states that utilization-focused evaluation is not oriented toward the evaluator making value judgments about the programs being evaluated, rather it is the evaluator's responsibility to train stakeholders to process and interpret the data. Based on the questions identified, the primary users may then make value judgments themselves.

Finally, evaluation reporting is intended to facilitate intended use by intended users. Again, the various primary stakeholders may participate in decisions about how the findings should be disseminated to potential user groups, as well as how the findings should be reported. Typically the preparation of the report or presentation of findings is prepared by the evaluator, guided by preferences of stakeholders.

Responsive Evaluation

Robert Stake has been at the forefront of evaluation theorists in proclaiming the necessity for involving stakeholders in evaluation. While Stake's name is associated with many important evaluation ideas, an examination of "responsive evaluation" may shed the most light on his views related to stakeholder participation. Stake notes that the evaluator's role is geared to the identification of issues of greatest importance to those most affected by the evaluation, typically practitioners. He emphasizes the importance of identifying issues based on stakeholder's multiple perceptions and constructed realities, noting that people may have "multiple purposes, observations and value judgments" (1991, p. 84).

Stakeholders are broadly defined in responsive evaluation. While Stake notes that the overriding goal of including all program stakeholders to ensure political fairness and justice in the evaluation, his more recent writings exhibit a narrowing of who represents a stakeholder. He notes that, based on an orientation toward instructional improvement, responsive evaluation focuses on issues prescribed by the immediate stakeholders or local people with a "stake" in the evaluation, such as teachers, local administrators, community leaders and other onsite program directors and staff (Stake, 1991). These stakeholders are major participants in nearly all aspects of the evaluation. With the evaluator, acting as a "conscious-ness-raising educator," the stakeholders may aid in deciding problems, questions, interventions, and methods.

Initial planning in responsive evaluation begins with the evaluator acquiring a comprehensive understanding of the evaluand based on observations, interviews of stakeholders, and examination of documents. The evaluator then works to con-struct issues around which the evaluation will be designed. These issues are informed partly by the evaluator's acquaintance with similar programs but mainly from on-site contacts and negotiations with stakeholders (1991). A few issues are chosen ultimately by the evaluator as the evaluation's conceptual organizers (Stake, 1991). Throughout the course of the evaluation, the list of issues is poten-tially adaptable due to the changing nature of the program.

While the evaluator engages in negotiation with stakeholders about alternative methodological approaches, the evaluator ultimately determines the method, based on: (1) the needs, resources, and valuing of the stakeholders; (2) a knowl-edge of different methodological applications; and (3) the skills of the evaluation team (Stake, 1991). The goals of responsive evaluation, however, tend to orient data collection toward naturalistic or qualitative methods to collect the data, such as multiple observation, interviews, and case studies.

Stakeholders are also readily involved in the data collection. Stake sees stake-holders as informants, not subjects, whose subjective perceptions are collected and triangulated with other sources to show the reliability of the observations. Because of his belief in multiple realities, the various stakeholders are encouraged to make value judgments about the program issues at hand. Stake notes that "part of the description, especially about the worth of the program, is revealed in how people subjectively perceive what is going on. Placing value on the program is not seen as separate from perceiving the program" (Stake, 1991, p. 79). Further, in responsive evaluation, the evaluator should reveal "minority value positions" (Stake, 1975, p. 37), that is the values of stakeholders whose numbers may be so small to not otherwise be heard in the evaluation.

In reporting findings of the evaluation, stakeholders may guide report writing and recommended outcomes. Evaluation findings are presented to facilitate stake-holder understanding and interpretation, leading the "local people to discover and construct their own truths, their own definitions of the problem, and their own solutions" (Shadish, Cook, & Leviton, 1991, p. 303). Ultimately, presentation of evaluation results occurs through a variety of modalities, depending on the stake-holders' needs and desires. This may include reporting to stakeholders in the natural language through a report, case study, or briefing panel. Through this approach, Stake believes that stakeholders assume greater ownership of the evaluation, and are more likely to use the evaluation findings for program and instructional improvement.

Participatory Evaluation

A number of evaluation theorists use the term participatory evaluation to describe their work (e.g., Greene, 1987, 1988a, 1988b, 1990). Others, while not specifically using the term, could easily have their work listed within this category (e.g., Fetterman, 1996). Cousins and Whitmore (1998) have delineated distinctions between forms of participatory evaluation.

We will focus first on describing the views of Cousins and Earl. According to these theorists, participatory evaluation focuses on enhancing evaluation utilization by increasing both the depth and the breadth (or range) of primary users' involvement in the evaluation process (Cousins & Earl, 1992). While Cousins and Earl regard participatory evaluation as an extension of stakeholder-based models, they argue that participatory evaluation differs from "conventional" stakeholder-based evaluations in three important ways: (a) engaging only a relatively small number of "primary users" (see Alkin, 1991); (b) deep stakeholder participation in a broad range of evaluation activities (not merely functioning in a consultative way, but involved in the "nuts and bolts" of evaluation; and (c) the evaluator role as coordinator with responsibility for technical support, training, and quality control.

Cousins and Earl (1992, p. 400) maintain that "[C]onducting the study is a joint responsibility (for both the evaluator and the primary users)." The evaluator in participatory evaluation provides key organizational personnel with technical training so that they develop sufficient technical knowledge and research skills. These program practitioners "learn on the job' under the relatively close supervision of the expert evaluator while both parties participate in the research process" (Cousins & Earl, 1996, p. 8). Such learning, they claim, is vital to participatory evaluation, because key organizational members are expected to develop sufficient technical knowledge and research skills so that they will be able to coordinate their continuing and new projects, and only consult the evaluator about such technical issues and tasks as statistical analysis, instrument design, technical reporting, and so forth.

While the participatory model attempts to engage the selected stakeholders in the "nuts and bolts" of problem formulation, instrument design or selection, data collection, analysis, interpretation, recommendations, and reporting, Cousins and his colleagues acknowledge that in most participatory evaluations "stakeholder participation is generally limited to evaluation tasks that are not heavily technical." However, they assert that the depth of participation was increased, since "stakeholders...participate in developing instruments and collecting data, tasks that are more deeply involved than merely informing the evaluation design and helping to interpret data" (Cousins, Donohue, & Bloom, 1996, p. 223).

Also falling within the broader notion of participatory evaluation is empowerment evaluation, championed by David Fetterman and others. According to Fetterman, this approach encourages the use of evaluative techniques to help people engage in self-evaluation and self-reflection to improve their own programs (1996). The evaluator assumes a secondary role as collaborator and facilitator, with stakeholders actively involved in all stages of the evaluation process. The stakeholders themselves are "responsible for conducting the evaluation. The group thus can serve as a check on individual members, moderating their various biases and agendas...Participants in an empowerment evaluation thus negotiate goals, strategies, documentation, and time lines" (Fetterman, 1996, p. 23).

Fourth Generation Evaluation

Guba and Lincoln's descriptions of fourth generation evaluation suggest two key elements that are embedded in this approach. The first feature is that fourth generation evaluation determines what information is needed based on the claims, concerns, and issues of stakeholders. The second distinctive characteristic lies in its use of constructivist methodology to implement the evaluation. (Guba & Lincoln, 1989)

Guba and Lincoln regard the empowerment and enfranchisement of stakeholders as the central task in fourth generation evaluation. They identified three groups of stakeholders (Guba & Lincoln, 1981). The first group is 'agents,' those who are "involved in producing, using, or implementing the evaluand [the entity to be evaluated];" The second group is 'beneficiaries', those who "profit in some way from the evaluand"; The third group is 'victims', those who "are negatively affected by the evaluand" (Guba & Lincoln, 1989, pp. 201-202).

How are the stakeholders involved in the evaluation process? As noted, one key element on which fourth generation evaluation rests is responsive focusing, organized around the claims, concerns, and issues of stakeholders. This means that the determination of what questions to ask and what information to collect is based on stakeholder inputs. Therefore, stakeholders' active involvement in initial problem identification is guaranteed in fourth generation evaluation.

Guba and Lincoln further claim that "[t]he involvement of stakeholders in fourth generation evaluation implies more than simply identifying them and finding out what their claims, concerns, and issues are" (Guba & Lincoln, 1989, p. 56). Stakeholders should also maintain their continuing involvement in problem identification. Guba and Lincoln insist that each stakeholder group should confront and consider the inputs from other groups. If these inputs conflict, then stakeholders must deal with them "either reconstructing their own constructions sufficiently to accommodate the differences or devising meaningful arguments as to why the others' propositions should not be entertained" (Guba & Lincoln, 1989, p. 56).

Fourth generation evaluation also treats stakeholders as equal partners in every aspect of design, implementation, interpretation, and resulting action of an evaluation. Guba and Lincoln argue that one advantage of the constructivist methodology is that it does not treat stakeholders' emic constructions—the values/beliefs held individually prior to the onset of evaluation—as biased perceptions, but as legitimate. It thereby enables stakeholders to control the evaluation activity.

Guba and Lincoln's descriptions of the way in which interpretations and reinterpretations of realities are achieved suggest that stakeholders in fourth generation evaluation can be a source of data for evaluation, as would be the case in many or most evaluations. The kind of data they provide, however, differs from data collected in a traditional positivist orientation of evaluation (e.g., participants' test scores). According to Guba and Lincoln, stakeholders, chosen for open-ended interviews have opportunity to voice their comments on the 'evaluand'. These comments constitute a data source and "might include observations about claims, concerns, and issues, and observations about what is liked and disliked about the evaluand" (1989, p. 151). The data collected, Guba and Lincoln maintain, are used to form the insider interpretation of the evaluand, which is the focus of the evaluation inquiry.

Guba and Lincoln also give high priority to stakeholders in making value judgments. They suggest that what evaluators would consider findings in fact should be negotiated between and among the various stakeholders, so that evaluators can check whether they got it right. Guba and Lincoln argue further that all stakeholders access to the 'story' evaluators constructed and should have the opportunity to tell evaluators of any misrepresentation. This way of addressing textual validity— a representation of which is accurate, true, and complete—has stakeholders validate or legitimate the stories that evaluators constructed as authentically theirs (Lincoln, 1994). In other words, what is to be considered true is left to negotiations between the evaluator and stakeholders.

The report of a fourth generation evaluation reflects the shared construction of both outsider and insider viewpoints. Guba and Lincoln think it is good for the evaluator to prepare reports "individually to each of the stakeholding audiences, in the spirit of empowerment..." (1989, p. 225). The evaluator should cooperate fully with stakeholders in presenting the report in language appropriate to that stakeholder group.

SOME CASE EXAMPLES

Often, the implementation of theory into practice suffers in its translation. Thus, in the examination of the following case examples, very few follow precisely in the manner prescribed by a particular model. Nonetheless it is instructive to examine the way in which stakeholders were involved in a number of specific case examples.

Evaluation of the Cities-in-Schools (CIS) Program

The NIE stakeholder approach was first attempted in the evaluation of two national programs; one of which was the Cities-in-Schools (CIS) program evaluated by American Institutes for Research (AIR). The CIS program was designed to coordinate certain youth services in a number of cities. This coordination was believed to have a good chance of "drawing the most intransigent young people into ordinary, desirable educational and social behaviors" (cited in Stake, 1983, p. 16). Examples of the treatments that the CIS program included: organizing students into forty-member "families," and providing these students with 24-hour-available tutors from four agency outreach persons, who "hung around" with these students. These treatments were believed to be able to produce such desirable outcomes as increased school attendance, improvement in achievement in basic skills, and avoidance of trouble with the police.

An evaluation of CIS programs was conducted in three of the cities: Atlanta, New York City, and Indianapolis. Charles Murray, the study director, stated that the emphasis in the CIS evaluation was to increase the use of evaluation findings. To achieve this goal, the stakeholder, using Murray's words, "(1) can tell the evaluator what to measure, (2) can tell the evaluator when and in what form the evaluation must be presented to be useful, [and] (3) will be more likely to pay attention to the evaluation results when they are presented" (cited in Stake, 1983, pp. 15-16).

Murray identified three different groups of stakeholders in CIS. One group was the program staff people. The second was the clients of the program, the parents and the students themselves. The third group was the decision makers, the funding agencies, and the people who would decide whether the program would be institutionalized. Murray further divided the third group into local and national stakeholders. Murray's definition of stakeholder, however, reveals that he viewed the stakeholder as an information user and not as an active participant. The stakeholder, Murray declared, is responsible for assuring that the information gathered will be useful.

How were stakeholders involved in the evaluation of CIS? Despite Murray's description of the multiple ways that stakeholders are to be involved, the stakeholder idea was only partially exploited in the CIS evaluation. Stakeholders were regularly asked four questions: (1) What indicators would convince you that the project is successful? (2) What are the most important of these indicators for you? (3) When do you need feedback from the evaluation? and (4) In what form would you prefer that feedback to be? From the first two questions, it can be seen that stakeholders were provided the opportunity to be involved in the identification of problems. Their information needs, however, were mainly identified to enable evaluators to know the stakeholders' desired outcome. Stakeholders, therefore, were contributors to the information matrix, but not potential collaborators in inquiry.

CIS staffers, a potentially important stakeholder group, did not participate in the above process. Their exchanges with evaluators were more for the purpose of providing data than for the purpose of serving their own information needs.

Feedback was to be provided to representatives of stakeholder groups and reports were to be made available for those who requested them. AIR's proposal suggests that stakeholders were mainly perceived as evaluation report audiences represented by a few individuals. While stakeholders could specify when and in what form they wanted the report, they did not participate in the preparation and write-up of the report, or in making value judgments about the relative importance of different findings. These tasks were provided by Murray.

The review of the CIS evaluation suggests that the idea of stakeholder evaluation as articulated by Gold was only partly exploited in the CIS study. Many factors attributed to this. One difficulty AIR encountered was that their alliances with program staff were weak. As a result, data collection suffered because the staff did not trust the evaluators enough. The weak relationship between the evaluators and stakeholders was partly due to the multiple stakeholder audiences involved in the study, the contractual obligation to a federal office, and the strong professional affiliation with the academic community. But the major underlying deficiency was AIR's lack of consideration of the possible difficulty in communicating with stakeholders or in understanding their needs. One of the key ideas in NIE's stakeholder approach was to reorder the evaluator's relationship with stakeholders so as to increase evaluation use. Yet the AIR proposal did not indicate the possible difficulty in communicating with stakeholders.

Another difficulty was caused by the broad spectrum of stakeholders and the failure of the evaluator to operationalize an appropriate procedure for involvement. Murray defined the users (or stakeholders) for CIS as representing "...an unusually broad range of individuals and groups. There are important stakeholders in the executive branch of the federal government and in state and local government agencies; in the community, there are stakeholders in social service agencies, commercial enterprises, educational institutions, in neighborhood and consumer groups" (cited in Stake, 1983, p.20). To meet the needs and requirements of such a wide range of stakeholder groups was 'foolhardy', using Murray's self appraisal.

The third difficulty was closely related to the second one: how to establish a stakeholder kind of relationship with such a diverse stakeholder population as defined above? Murray acknowledged, "The idea of using clients as stakeholders never really got off the ground. How do you get parents and kids together in Indianapolis?... [In the end, [y]ou get parents who really like the program....But this isn't having a stake in it. I don't see how you establish a stakeholder kind of relationship here" (cited in Stake, 1983, p. 26). As mentioned previously, the meeting with stakeholders representing national decision makers was equally unpromising. As Murray said, "Many were at the meeting only because they had been ordered there by cabinet secretaries" (cited in Stake, 1983, p. 22). With respect to the meetings with local decision makers, one common response to the issue of 'what constitutes success for you' was that "Listen, you folks are being paid lots of money to do this [evaluation]. You're supposed to be experts on what constitutes success [for us]" (cited in Stake, 1983, p. 22). This might have to do with the fact that the question probably was not skillfully asked; on the other hand, it might be due to difficulties in establishing a stakeholder kind of relationship in a large-scale evaluation and that the evaluator has to devise adaptations that engage the broad spectrum of stakeholders.

Examination of the particular difficulties encountered in the NIE stakeholder evaluation of CIS suggests several things. On the one hand, it might be appropriate to question whether or not the stakeholder approach is appropriate for a national program which consists of such complex program structures and involves such diverse program-concerned groups. Further, it is apparent that the possibility of a stakeholder approach being successful requires an understanding of the approach and, most importantly, a full commitment to its implementation by the evaluator.

Evaluation of The Caribbean
Agricultural Extension Project (CAEP)

An example of stakeholder involvement in a utilization-focused is presented in the Caribbean Agricultural Extension Project (CAEP) (Alkin, Adams, Cuthbert, & West, 1984). The purpose of the evaluation, funded by the U.S. Agency for International Development (AID), was to increase the effectiveness of agricultural extension services in eight Caribbean countries (Alkin et al., 1984, p. 2).

The key stakeholders were: (1) U.S. AID Caribbean office, especially the agricultural staff; (2) University of the West Indies Project staff and faculty in agriculture; (3) agriculture and extension officials in the eight participating countries; (4) farmers and their representatives on the project advisory committee; and (5) the American academic staff involved in the project, including Michael Patton as project director.

The evaluation team was carefully chosen to represent various primary stakeholders on several bases: disciplinary diversity; gender balance and expertise in various project aspects (agriculture, communications, higher education, vocational education). The team consisted of four members: an agricultural economist at the University of Missouri, chosen by U.S. AID; a University of the West Indies (UWI) faculty member in the communications arena; an Ohio State University faculty member, chosen by the American academic participants, and a UCLA faculty member experienced in utilization-focused evaluation, chosen as team leader.

In particular, the selection of one team member by a key stakeholder (UWI staff) was critical. Not only was the team member known as trustworthy, but "her perspective was necessary to make the whole process credible to UWI staff. Her presence was an assurance that the evaluation wasn't something that was being controlled by the outside, because she was an 'inside' Caribbean person. That was important to the team makeup and its wider acceptability" (Alkin & Patton, 1987, p. 21). Likewise, the selection of the agricultural economist by U.S. AID assured this stakeholder group that important technical concerns would be addressed. Thus, the presence of both an "insider" and "expert" outsiders, as well as the evaluation team's work over time, added to their overall credibility as an "evaluation team" (Alkin & Patton, 1987).

The two-year CAEP evaluation design collected needs assessment and planning data from eight Caribbean countries (Antigua, Belize, Dominica, Grenada, Montserrat, St. Kitts/Nevis, St. Lucia, and St. Vincent), with an advisory committee in each country. Information was collected at the primary training institution (University of West Indies, Trinidad) and at various agencies in both Trinidad and Barbados. Data collection included interviews (using detailed protocols, with input from various stakeholders during the development process), document reviews, attendance at meetings, as well as individual country summaries by evaluation topic. Additionally, the evaluation team conducted a series of case studies among farmers to document program impacts.

Concern for stakeholder involvement was paramount throughout the evaluation process. Not only were stakeholder interests represented in the selection of the evaluation team, but the CAEP project director stressed the credibility, competence, and representativeness of the team to the other stakeholders throughout the evaluation. Further, the primary intended users (U.S. AID, UWI, etc.) were able to meet and improve the evaluation team prior to hiring.

The evaluation began with a three-day regional advisory meeting of all the stakeholders (about 50 participants total) in April 1983, including the representatives from each of the eight participating countries and the other stakeholder organizations involved in agricultural development in the Caribbean. During this meeting, the evaluation team focused on determining issues of interest to the various primary intended users, rather than preconceived external criteria for what agricultural extension programs were supposed to accomplish.

Based on input from the initial meeting, the evaluators focused the draft design on relevant criteria, based primarily on the user questions at hand. They ultimately created a detailed, clearly-defined, user-driven evaluation design. This draft design report showed what the data sources were going to be and gave a preview of the evaluation report's format and content based on participant reported summaries of meeting recommendations. The team leader met with U.S. AID to discuss their impressions and all other primary users reviewed the design prior to beginning the field work.

Because the stakeholders had contributed to the evaluation design, the evaluation team was not perceived by the stakeholders as gathering data extraneous to the evaluation issues and questions at hand. The design was readily adopted by program staff who used it as a basis for the project workplan, and subsequently as a part of their internal quarterly reporting system. Monthly staff meetings were held, structured around the evaluation's elements, to discuss the latest activities for the major outcome categories. This approach provided constant feedback and accountability measures to ensure systematic progress, thus enhancing clear communication between the stakeholders and evaluation team members. As the evaluation findings were so relevant, any concerns about the evaluation were shortlived. Alkin and Patton (1987, p. 23) noted the importance of the evaluation-derived work plan as a focusing tool for the entire project, stating: "The evaluation design and questions, in essence, became the structure for project self-monitoring."

Maintaining close user-relationships (as is typical in utilization focused evaluation) posed some difficulties to the evaluators. The primary concerns were the distance between the data collection sites, unavailability of telephones in some locations, heavy traveling, and other logistical problems. Because of the geographic spread, the evaluators were unable to engage in informal, spontaneous interactions and rapport-building with stakeholders that frequently characterize utilization-focused evaluations. To the extent possible, however, the evaluation team tried to remedy any stakeholder concerns by building a clearly defined and organized evaluation plan, the relation between each of the steps, and the procedure to be followed. Alkin states:

> The procedure was incredibly detailed and might have been viewed as overly burdensome for most evaluations. In this particular evaluation, however, the evaluators felt a need to make sure that a lot of these steps were visible, as a means of maintaining credibility. That is, there was a clear and well connected sequence of evaluation activities, from design, evaluation questions, data, and instruments. And the important stakeholders 'bought into' the final results because the sequence was so clear, and they had reviewed and approved it at a number of junctures along the way (Alkin & Patton, 1987, p. 24).

Since frequent communication and informal feedback took place between the stakeholders and evaluation team members and draft versions of the report were reviewed by project staff and U.S. AID personnel, some stakeholders were aware of details of the report prior to publication. The report showed substantial progress and major successes throughout the region. Further, because the stakeholders were so involved throughout the evaluative process, the evaluation had a "major impact on program implementation" (Alkin & Patton, 1987). It resulted in re-focusing the program implementation to identify and correct weaknesses, as well as to more directly focus on areas that were being neglected. Finally the evaluation led U.S. AID to continue project funding, an issue of dispute prior to the evaluation.

Evaluation of the Transition Years Initiative

Using the participatory approach, Cousins reports on evaluating a pilot program—Transition Years initiative—implemented in Rockland school district, east-central Ontario. (Cousins, 1997) The program was designed to better engage ninth grade students and foster their participation in school life, thereby reducing the dropout rate. At the beginning of the study, the evaluator met with the steering committee and together they decided the nature and scope of the study, the kind of data to collect, and the data collection methodology. The steering committee consisted of two superintendents, one principal, one vice-principal, one district curriculum coordinator, and four teachers added later following the evaluator's suggestion. A training session on interview procedures followed. Upon completion of data collection, members of the research team were trained to independently code their data and then met to integrate findings. A final report was drafted by the evaluator and revised by the teacher teams before it went forward to the steering committee. At the end of the study, the evaluator met with the steering committee to discuss the report, identify changes, and plan dissemination and follow-up activities.

The Rockland evaluation results were disseminated within the district and recommendations were reported as well as adopted. Follow-up research activities were carried out by the stakeholders (steering committee), without the evaluator's further involvement due, in large part, to the training provided.

Reviewing his experience in conducting the participatory evaluation, Cousins drew three conclusions about the depth of primary users involvement. First, it is better not to involve participants in highly quantitative data analysis, although it is important to train them to be able to make sense of the statistical output. Second, it is worthwhile to engage participants in interview data collection and the content analysis of qualitative data, although these activities are labor intensive. And last, primary users' participation in dissemination is vitally important.

Cousins' participatory evaluation defines primary users as "those who have the organizational authority and influence to act on findings and who are connected to program management and direction" (Cousins et al., 1996, p. 223). An examination of the Rockland case, however, suggests that there were two different stakeholder groups involved in this evaluation. The steering committee (the primary users by his definition) engaged primarily in planning, instrument development, and interpretation of findings. On the other hand, the teacher teams were involved mainly in data collection, qualitative data analysis, interpretation of findings, and write-up of report. Thus, in the case example participatory evaluation allows for different levels of stakeholder participation and responsibility: primary users and a stakeholder group without organizational authority to act on findings.

Evaluation of the Youth Employment Program

Stakeholder participation is linked closely with evaluation utilization in the evaluation of the Youth Employment program (Greene, 1988). The purpose of the program was to help youth, ages 14 to 19, gain meaningful work experience through placement in a summer job. In this evaluation, both individual and groups of stakeholders participated in all phases.

Initial discussions between the researcher/evaluator occurred with three main figures—the coordinator of the Youth Employment program, the head of the Youth Development department, and the agency director. This was followed by numerous discussions with approximately 15 stakeholders to develop an evaluation plan, including the substantive focus and a specific evaluation question (what are private sector employers' perspectives on youth employment and the congruence between employer and youth job-related needs and expectations) and an evaluation design (purpose, audiences, methods, and expected uses of information for program improvement). Stakeholders were selected for "maximum diversity and representativeness" (Greene 1988, p. 93), and included program funders, other youth professionals, employers, and youth, agency board members, administrators, and program staff.

Data for the evaluation included individual interviews with the evaluator, group meetings with other stakeholders, and written responses to mailed questionnaires. According to Greene (1988, p. 93), "These varied vehicles for participation were iteratively structured first to generate a wide, divergent set of program concerns and issues, reflecting multiple perspectives and values, and then to come to agreement on the concerns of highest priority." Additional data were collected through individual and group interviews with different kinds of employers.

Evaluators ensured that all stakeholders were consistently updated on the progress of the evaluation, invited their participation in collecting data and in informing data collection activities, and encouraged their interpretation and analysis of the data. More intensive communication occurred with a subgroup of stakeholders (four youth bureau staff), who regularly informed questionnaire development. This communication occurred largely through numerous progress reports of completed activities and upcoming activities, as well as through consultative meetings with the evaluator. When data were available, nontechnical narratives (descriptive results, relational results) were shared with stakeholders, then discussed in meetings with either the evaluation subgroup or all the stakeholders.

Stakeholders reflected on and interpreted the results and then identified additional questions and ways to examine them. Additionally, subgroups of stakeholders (e.g., agency board, Youth Development board subcommittee) reviewed the interim written summaries, discussed the potential impact of the evaluation, and ensured that the implications and recommendations were appropriate, meaningful, and feasible. Feedback and recommendations (in elaborated program scenarios) from the subgroup meetings were incorporated into the summary prior to review with the following subgroup. Ultimately, a final written evaluation report was prepared, bound, and disseminated to all stakeholders who participated in the evaluation. According to Greene (1988), the evaluation had immediate useful effects. Comments following the evaluation suggest that the evaluation was helpful for increasing visibility of the Youth Employment program, aiding in justifying budget requests and enhancing program staff expansion.

LESSONS LEARNED FROM STAKEHOLDER APPROACHES

There are a number of patterns that can be discerned from examining the various stakeholder oriented evaluation approaches and their application. The stakeholder concept also leads to a number of issues yet to be resolved. In this concluding section, we will examine the following areas: (1) the differential impact of evaluation theoretic approaches on the depth and breadth of stakeholder participation; (2) the essential role of the evaluator on the successful inclusion of stakeholders; (3) selection of stakeholders; and (4) the "impact" of stakeholder involvement.

Similarities and Differences Related to Evaluation Approach

It is apparent from the previous discussion that all of the evaluation approaches described have provisions for the evaluator to encourage and support stakeholder participation throughout the evaluation process. In theory, at least all of the views presented were committed to systematically engaging stakeholders in defining the purpose and the objectives of the evaluation, providing input into instruments and data to be collected, and participating in defining the character and format of final reporting. Some approaches went well beyond this by also engaging stakeholders in such things as the development of instrumentation, collection of data, and the valuing of findings. Furthermore, approaches differed in the number of stakeholders viewed as appropriate for participation.

The most clear cut example of differences in the range of stakeholder activities (the depth of stakeholder involvement) focuses on whether an approach is trying to be responsive to stakeholders' needs, values, and opinions (e.g., NIE Stakeholder approach) or is oriented to letting stakeholders control the evaluation with the evaluator as facilitator (e.g., participatory evaluation). These two examples may be considered as illustrative, perhaps, of a continuum of views. Clearly we have noted gradations in the manner in which evaluators attempt to be responsive to stakeholders' needs, values, and opinions. Is this, for example, a one-time event or is there a continuing relationship between evaluator and stakeholders? And, does the relationship vary by evaluation purpose?

The various approaches also differed in the breadth of stakeholder involvement. While some approaches (e.g., utilization-focused evaluation and participatory evaluation) are oriented toward a limited set of stakeholders (primary users), others define stakeholder audiences much more broadly (e.g., fourth generation evaluation).

These differences in the breadth and depth of stakeholder involvement provide for substantial differences in evaluation theoretic approaches. Thus, as initially noted, there is no single stakeholder evaluation "model."

The Evaluator's Role

As has been noted in the literature, the evaluator's role and commitment to a theoretic point of view is pivotal in the successful implementation of evaluations (Alkin et al., 1979). Aside from the nature of the evaluation theory (breadth and depth of stakeholder participation) there is the necessity for an evaluator to be committed to the implementation of that approach. As noted in the CIS Evaluation, the failure of the evaluator to be committed to having stakeholder participation led to a very poor application of the NIE stakeholder approach. Moreover, demonstrated within the same case was the necessity for the evaluator to have the interpersonal and political skills necessary to maximize stakeholder participation, within the constraints of the particular program being evaluated. To be successful in implementing evaluations with stakeholder participation, evaluators must be able to free themselves of the constraints of prescribed evaluation procedures and to be situation adaptable. Evaluators who are both responsive and flexible are thus best suited for engaging stakeholders in the evaluative process.

Selection of Stakeholders

The most prevalent technical issue emerging from the literature and demonstrated to some extent within the case studies is whether a single evaluation can comprehend and represent the interests and views of many different stakeholders. This is dealt with to a large extent by those evaluation theoreticians who chose to focus on a narrower set of stakeholders for inclusion in the evaluation (e.g., UFE, participatory evaluation). In large part, this issue is associated with the particular value orientation of the theoretic approach employed and the way in which it defines stakeholders.

An associated issue relates to multi-governmental level programs and other programs offering similar complexity. Again, this was particularly evident in the CIS Evaluation which required, but did not have fulfilled, an adaptation on the part of the evaluator to obtain appropriate stakeholder participation.

The power to select stakeholders is an awesome responsibility—it goes to the very root of the value assumptions of evaluators. The issue is more than which stakeholders will participate but rather *which* stakeholder voices will be heard. As we have noted the nature of that decision for different evaluators rests on their view of the purpose of evaluation and the attendant role of the evaluator. Irrespective of the evaluators' views it is necessary for evaluators to be ever mindful of the possibility of being co-opted by particular stakeholders or stakeholder groups so that their evaluations do not properly reflect stakeholder priorities, in accordance with their own theoretic perspective.

The "Impact" of Stakeholder Involvement

With respect to the issue of the existence of "demonstrable" evidence that stakeholder involvement makes a difference, the evidence would appear to be positive. While there is an absence of controlled studies demonstrating the relative benefits of a stakeholder evaluation versus a similar evaluation without stakeholder involvement there is other convincing data. Patton (1988, p. 325) notes "the massive diffusion of innovation literature...[and] the formal organizations, participatory management and small group literature's in psychology and sociology [which] provide substantial data relevant to this point. These literature document with empirical evidence the proposition that people are more likely to accept and use information, and make changes based on information, when they are personally involved in and have a personal stake in the decision making processes aimed at bringing about change."

There has been a recent evaluation literature on the emergence of what Patton (1997) calls "process use." By this, he means the stakeholders' development of skills through their engagement in the process of evaluation. Patton maintains that, in many respects, this kind of process-oriented use is often more powerful than is the actual use of results.

Theoretic Concerns Related to Stakeholders

A number of concerns have been raised related to the way various evaluation theoretic perspectives engage stakeholders. One debate revolves around which if any, stakeholders' interests should guide the evaluation. For example, the focus on primary intended users, frequently the client, found utilization-focused evaluation and in participatory evaluation has been criticized by other major theorists. House is critical based upon social justice criteria—the failure to serve underrepresented populations (Alkin, 1990). Scriven, although interested in serving client needs, has generally viewed all approaches which attend directly to client interests as potential usurpation by clients.

Another issue is the extent to which the evaluator can adequately represent the interests of stakeholders. In responsive evaluation, for example, the evaluator interviews primary stakeholders and synthesizes their views to determine the focus of the evaluation. Such a notion, however, may be difficult to translate into practice.

Finally, "advocacy" is another issue of concern related to stakeholder participation in various evaluation approaches. For example, in empowerment evaluation, evaluation credibility is potentially undermined because the evaluator becomes an advocate for the group (oppressed people) rather than an advocate for the evaluation findings. Likewise, advocacy is an issue in fourth generation evaluation. Evaluators of this theoretic persuasion advocate the right of the various stakeholders to have their different perspectives; the perspectives become the ends rather than the means. The intent is to have the different perspectives engage each other.

FINAL NOTE

A number of evaluation theoretic points of view have been presented to demonstrate the way in which stakeholder concepts pervade evaluation. Clearly, there are a multiplicity of activities in which stakeholders are engaged in evaluation, as well as divergent opinions regarding the impact of their participation. These differences have been delineated related to various evaluation theoretic views. Nonetheless, despite criticisms and challenges related to stakeholder participation, the perceived benefits of this engagement remain high.

ACKNOWLEDGMENT

We wish to acknowledge the helpful suggestions to earlier drafts made by Brad Cousins, Ernie House, John Owen and Mike Patton. The authors assume responsibility for any errors or misinterpretations.

REFERENCES

Alkin, M.C., Daillak, R., & White, P. (1979). *Using evaluations: Does evaluation make a difference?* Beverly Hills, CA: Sage Publications.

Alkin, M.C., Adams, K.A., Cuthbert, M., & West, J.G. (1984). *External evaluation report of the Caribbean Agricultureal Extension Project, Phase II.* Minneapolis: CAEP.

Alkin, M.C. & Patton, M.Q. (1987). Working both sides of the street: The client perspective in evaluation. *New Directions for Program Evaluation, 36,* 19-32.

Alkin, M.C. (1990). *Debates on evaluation.* Newbury Park, CA: Sage Publications.

Alkin, M.C. (1991). Evaluation theory development. In M.W. McLaughlin & D.C. Phillips (Eds.) *Evaluation and education: At quarter century, Part II.* (pp. 89-112). Chicago: University of Chicago Press. 89-112.

Cousins, J.B. & Earl, L.M. (1992). The case for participatory evaluation. *Educational Evaluation and Policy Analysis, 14*(4), 397-418.

Cousins, J. B. & Earl, L. M. (1995). *Participatory evaluation in education: Studies in evaluation use and organizational learning.* London: The Falmer Press.

Cousins, J. B., Donohue, J. J., & Bloom, G. A. (1996). Collaborative evaluation in North America: Evaluators' self-reported opinions, practices and consequences. *Evaluation Practice, 17*(3), 207-226.

Cousins, J.B. (1997). Organizational consequences of participatory evaluation. In K.A. Leithwood & K.S. Louis (Eds.), *Communities of learning and learning schools: New directions for school reform.* Amsterdam: Swets & Zeitlinger.

Cousins, J.B. & Whitmore, E. (in press). Conceptualizing participatory evaluation. In E. Whitmore (Ed.). *New Directions For Program Evaluation.*

Fetterman, D.M. (1996). Empowerment evaluation: An introduction to theory and practice. In D.M. Fetterman, S.J. Kaftarian, & A.Wandersman (Eds.), *Empowerment evaluation: Knowledge and tools for self-assessment and accountability.* Thousand Oaks, CA: Sage Publications.

Gold, N. (1983). Stakeholders and program evaluation: Characterizations and reflections. *New Directions for Program Evaluation, 17,* 63 72.

Greene, J.C. (1986). *Participatory evaluation and the evaluation of social programs: Lessons learned from the field.* Presented at Annual Meeting of the American Educational Researcher Association. San Francisco, CA. (ERIC Document Reproduction Service No. 291 766).

Greene, J. C. (1987). Stakeholder participation in evaluation design: Is it worth the effort? *Evaluation and Program Planning, 10,* 375-394.

Greene, J.C. (1988a). Communication of results and utilization in participatory program evaluation. *Evaluation and Program Planning, 11,* 341-351.

Greene, J.C. (1988b). Stakeholder participation and utilization in program evaluation. *Evaluation Review, 12,* 91-116.

Greene, J.C. (1990). Technical quality versus user responsiveness in evaluation practice. *Evaluation and Program Planning, 13,* 267-274.

Guba, E.G. (1969). The failure of educational evaluation. *Educational Technology, 9*(5), 29-38.

Guba, E. G. & Lincoln ,Y.S. (1981). *Effective evaluation.* San Francisco: Jossey-Bass.

Guba, E. G. & Lincoln, Y. S. (1989). *Fourth generation evaluation.* Newbury Park, CA: Sage Publication Inc.

House, E. (1991). Evaluation and social justice: Where are we? In M.W. McLaughlin & D.C. Phillips (Eds.) *Evaluation and education: At quarter century, Part II.* (pp.232-245). Chicago: University of Chicago Press.

House, E. (1993). *Professional evaluation.* Newbury Park, CA: Sage Publications.

House, E. (1995). Putting things together coherently: logic and justice. *New Directions for Program Evaluation, 68,* 33-48.

Lincoln, Y.S. (1994). Tracks toward a postmodern politics of evaluation. *Evaluation Practice, 15*(3), 299-309.

Mark, M.M. & Shotland, L.R. (1985). Stakeholder-based evaluation and value judgments. *Evaluation Review, 9*(5), 605-626.

Patton, M.Q. (1986). *Utilization-focused evaluation, 2nd edition.* Newbury Park, CA: Sage Publications.

Patton, M.Q. (1988). Six honest men for evaluation. *Studies in Education Evaluation, 14*(3), 301-330.

Patton, M.Q. (1997). *Utilization-focused evaluation: The new century text, edition 3.* Thousand Oaks, CA: Sage Publications .

Rossi, P.H. & Freeman, H.E. (1993). *Evaluation.* Newbury Park, CA: Sage Publications.

Scriven, M. (1993). The nature of evaluation. *New Directions for Program Evaluation, 58,* 5-48.

Sechrest, L., Babcock, J., & Smith, B. (1993). An invitation to methodological pluralism. *Evaluation Practice, 14*(3), 227-235.

Shadish, W.R., Jr., Cook, T.D., & Leviton, L.C. (1991). *Foundations of program evaluation: Theories of practice.* Newbury Park, CA: Sage Publications.

Stake, R.E. (1975). *Program evaluation, particularly responsive evaluation.* Kalamazoo, MI: Western Michigan University Evaluation Center.

Stake, R.E. (1983). Stakeholder influence in the evaluation of Cities-in-Schools. *New Directions for Program Evaluation, 17,* 15-30.

Stake, R.E. (1991). Retrospective on "the countenance of educational evaluation." In M.W. McLaughlin & D.C. Phillips (Eds.), Evaluation and education: At quarter century: Part II, Chicago, IL: University of Chicago Press.

Weiss, C. H. (1983a). The stakeholder approach to evaluation: Origins and promise. *New Directions for Program Evaluation, 17,* 3-14.

Weiss, C.H. (1983b). Toward the future of stakeholder approaches in evaluation. *New Directions for Program Evaluation, 17,* 83-96.

PART II

IMPLEMENTATION PROCESS AND CONTEXTS

Chapter 6

EVALUATING IMPLEMENTATION

Ann R. McCoy and Arthur J. Reynolds

INTRODUCTION

Evaluations are conducted to determine the appropriateness of the theoretical underpinnings of a program's design, whether a program is being implemented in accordance with its design, or to assess the impact and utility of an intervention. In this chapter our focus is evaluating program implementation. Historically, implementation evaluation has been a frequently overlooked evaluation activity. Yet it is an essential part of the evaluation process. We review what implementation evaluation is, why it is important, when it is necessary, and how the process of implementation is measured for evaluation purposes. The role of evaluation of implementation in program development and analysis is illustrated for the Head Start preschool program and the School Development Program, educational interventions designed to promote successful educational and social outcomes. Finally, we discuss the implications of evaluating program implementation for program improvement and dissemination. In this chapter, no distinction is made between implementation and "process" evaluations. We use the terms interchangeably, though for clarity, implementation evaluation is often preferred.

Advances in Educational Productivity, Volume 7, pages 117-133.
Copyright © 1998 by JAI Press Inc.
All rights of reproduction in any form reserved.
ISBN: 0-7623-0253-4

WHAT IS IMPLEMENTATION EVALUATION?

Evaluating the implementation of educational and social programs is increasingly viewed as one of the most critical of all evaluation activities. Although its importance has been noted since the beginning of the evaluation profession (Charters & Jones, 1973; Weiss, 1972), the literature has expanded fairly rapidly in recent years (McGraw et al., 1994; Holden & Reynolds, 1997; Rossi & Freeman, 1993; Scheirer, 1994, 1996). This trend is timely in an age of rapidly changing educational and human-service programs in which coordinated and integrated services are high priorities. These new programs often are complex, and their components are valuable to study in their own right let alone as part of an outcome evaluation. Today's programs are also more likely to be decentralized. Site-to-site variations in program delivery can be large and these must be documented by evaluators.

Scheirer (1994) defines process evaluation as "the use of empirical data to assess the delivery of programs...[it] verifies what the program is, and whether or not it is delivered as intended to the targeted recipients and in the intended 'dosage'" (p. 401). Process evaluation is a means of documenting the chain of events involved in the implementation of an intervention. The documentation collected as part of a process evaluation details what the components of the intervention program were, how they were delivered and, who received them. Information obtained through process evaluation facilitates program dissemination thereby encouraging its future use by others (McGraw, Stone, Osganian, Elder, Perry, Johnson, Parcel, Webber, & Luepker, 1994). According to Holden and Reynolds (1997), process evaluations typically address three related questions:

1. Are the administrative services and resources that are key to the success of the program in place upon implementation?
2. To what extent are program services being delivered to the target population as specified in the program design?
3. What is the variation in program services and administration across program sites, and how do these relate to differences in the delivery and coverage of intended program services?

Implementation evaluation can be confused with needs assessment, evaluability assessment, and even some aspects of outcome evaluation but both the timing and purpose of process evaluation distinguishes it from each of these. Process evaluation occurs as a program is being implemented. After problems in a community are identified, needs assessments are conducted to measure their magnitude in order to devise programs and interventions to ameliorate these social ills (Rossi & Freeman, 1993). An evaluability assessment is conducted prior to the actual implementation or outcome evaluation to identify elements of a program's implementation which may significantly influence its effectiveness or to reach consensus about how best to conduct an evaluation. Impact evaluations are conducted after programs are implemented to determine the degree to which a program produced the intended effects. Process evaluation is concerned with evaluating the effectiveness of program implementation and it is usually more comprehensive than both monitoring and auditing.

Ultimately, the information collected as part of a process evaluation specifies a program's parameters and explains its effects. Program specification clarifies the program's intended purpose by linking data about a program's components with the outcomes it produces (Scheirer, 1994). When specifying a program, program developers and evaluators define which observable behavior will be investigated; create components which are discrete activities; state the theoretical rationale for each component; detail all components of an intervention even if they will not be measured; and indicate which components of an intervention can be tailored to meet the needs of the site as well as those which must be delivered as specified in a program's original design.

Precision in the specification of a program does not guarantee that its implementation will follow the prescribed course. Whereas program delivery may vary from site to site, the implementation of an intervention is subject to evaluation on two levels, macro and micro. Macro-implementation analysis tracks the interactions between and within intergovernmental agencies and the local bureaucracies responsible for providing the services they administer (e.g., state legislatures, state boards of education, school districts, schools and classrooms). Utilization of micro-implementation analysis within each of these settings will provide information on both the decision making processes and the level of support for or commitment to an intervention program among those responsible for its delivery. Decisions made at either the macro or micro-level can alter both the implementation of an intervention and its outcomes. Identification of the multiple points at which implementation problems can occur permits testing the effects of such problems on expected outcomes (Scheirer, 1994).

WHY IS IMPLEMENTATION EVALUATION A KEY ACTIVITY?

The investigation of program implementation is among the most important evaluation activities for three major reasons. We discuss these below (see also Holden & Reynolds, 1997).

To Assess a Fundamental Assumption of Impact Evaluation

Because a key assumption of outcome evaluation is that the program has been implemented as intended, a process evaluation allows a test of the accuracy of this basic assumption. Much previous research has indicated substantial deviations between the intended and actual implementation of treatments (Weiss, 1987; Blalock, 1980). Without knowing the extent and quality of participants' program experiences, inferences about the effects of the program are difficult to make with confidence. As Charters and Jones (1973) indicated, implementation evaluation is necessary to avoid studying "non-events." Committing such mistakes are called Type III errors. Process evaluation protects against a Type III error by demonstrating that the observed outcomes were produced by activities instituted as part of an intervention program and not other contemporaneous events. Providing a detailed picture of a program enables evaluators to determine the degree of fidelity between the designed program and the program as delivered; identify both the target population to be served by an intervention and the extent of services provided to that population; and to indicate how components of a program are linked to desired as well as undesired outcomes.

Unfortunately, problems in implementation often lead to the underestimation of the effect of the program but process analyses can suggest by how much. A related problem is that even if resources are available and services delivered, they may be delivered nonuniformly and not get to the target population most in need, as has often been observed for educational programs, especially large-scale block-grant programs like Title I (Chapter I) (Doernberger & Zigler, 1993).

To Help Explain Why a Program Worked or Did Not Work

Evaluating implementation helps researchers clarify how well the program elements were implemented well or not so well. This is particularly true for programs having multiple components. An assessment of the components that are well implemented can help determine the source of program effects. Alternatively, assessing implementation helps explain why the program may not have been as effective as expected or why effects varied across sites. There are three reasons why programs do not show their intended effects: (a) inadequate program theory or design, (b) poor program implementation, and (c) inadequacies in the research design or measures. Implementation evaluations help distinguish among these general explanations for smaller-than-expected findings. Without data on program implementation, evaluators cannot fully determine what may have happened to reduce effectiveness.

By identifying how particular intervention components contribute to achieving program outcomes, process evaluation strengthens the confidence in claims that a particular program or intervention caused the observed outcome. This makes it easier for policymakers to determine which programs are most likely to yield the best dividends for program participants and the larger community. For example, in explaining the effects of participation in an activity-based science program, Reynolds (1991) found that students' scores on a test of science achievement increased as a function of the number of science experiments their classes completed (a measure of program implementation) even after taking pretest scores and other factors into account. Such a dosage-response relationship strengthened the conclusion that the program was effective in raising student skill levels. Better program implementation may have resulted in even larger effects. Wide variation occurred in the number of experiments completed by students (2 to 12), the causes of which are another important question for implementation evaluation.

To Promote Program Replication and Evaluation Utilization

A third reason for the importance of implementation evaluation is that by understanding and documenting the workings of a program, the essential operative features of the program can be disseminated for use in other settings. Thus, implementation evaluation promotes program diffusion and replication. The dissemination of successful programs is facilitated by identifying actions organizations should engage in as well as those they should refrain from in order to achieve similar outcomes in different settings. In their evaluation of a diversion program for juvenile offenders called the Adolescent Diversion Project, Davidson and Redner (1988), for example, used this strategy in replicating the most important components of the original program in other settings. A complementary advantage of implementation evaluation is that it encourages collaboration between evaluators, program designers, and other stakeholders in understanding the essential features of programs that promote effectiveness. This also can lead to better utilization of findings. In the Adolescent Diversion Project, the most effective (and satisfactory to staff) program configuration for preventing recidivism included the components of diversion from adjudication, behavioral contracting, child and family advocacy, and relationship development rather than any one or two of these alone.

WHEN IS IMPLEMENTATION (PROCESS) EVALUATION NECESSARY?

An evaluation of program implementation should be conducted whenever there is a question concerning the match between the intended and actual operations of the program. In other words, such evaluations address the question "how does the program work?" (McGraw et al., 1996). Typically, programs or their operation are specified in one of three ways, explicating a program's underlying theory, formative evaluation, or evaluability assessment.

Specification of a program's underlying theory facilitates process evaluation by explicitly stating the process by which the program is expected to yield its intended results. According to Chen (1990), there are two major types of theory, normative theory (i.e., what stakeholders think the program should be and how it should work) and causal theory (i.e., identifying the linkages that connect interventions and outcomes based on current scientific literature). While these two types of theory should converge when developing and implementing a program often they do not. Process evaluation is useful in assessing such discrepancies. When a program's underlying theory is made explicit stakeholders and program evaluators benefit in the following ways: (1) those involved can articulate their assumptions about human behavior and how those assumptions influence their expectations about how the intervention will work and why; (2) program activities are connected to outcomes by identifying both components of a program's delivery and the necessary outcome measures; and (3) it clarifies the process by, explaining what is expected to happen and how, and establishes a cycle whereby theory is continuously articulated, tested and refined (Scheirer, 1994; Wholey, 1987).

Formative evaluation involves the assessment of a program's feasibility based on pilot data obtained during its development. This permits early identification and correction of problems associated with the delivery of a program before it has been codified and disseminated on a large scale. Formative evaluations offer an opportunity to preview a program and investigate the degree to which the theorized program is realized upon implementation. Discrepancies between what was expected and what transpired may serve to identify not only those aspects of the intervention which need to be altered but also those which must be eliminated in order to maintain the utility of the program.

Although these are distinct evaluation activities, process evaluations may also occur as part of evaluability assessments. Programs are often established as a result of an administrative order/ruling which does not specify a program's description or its underlying theory. As a result, the "same" program may be implemented at multiple sites while bearing only nominal similarities. Evaluability assessments are undertaken with existing programs to aid in determining what the program is meant to be and whether it is appropriate for evaluation before a full-scale evaluation takes place. The major stages in evaluability assessment entail the following: (1) clarifying the expectations of policymakers, education administrators, teachers, and school personnel regarding the program and its evaluation (to the extent that all stakeholders do not have the same expectations of or investment in a program its implementation may vary tremendously leading to variations in outcomes); (2) stakeholders describing how they expect providing services to result in the anticipated outcomes; (3) examining the program's theory in light of the actual functioning of the program; and (4) inviting the participation of policymakers and education administrators when determining how information obtained through an evaluation will be used (Scheirer, 1994).

How is Process Measured for Evaluation Purposes?

By definition a process is fluid, dynamic and constantly changing. This suggests that its measurement requires a high degree of detail in order to identify the micro-processes involved and illustrate how they combine to produce the achieved outcomes. Implementation data provides the most accurate means of assessing what actually happens during an intervention. This data allows evaluators to determine the following:

- extent of coverage and appropriateness of the target population
- amount of exposure to the program and its components provided to or experienced by the participants
- fidelity with which the program was delivered as intended and the level of satisfaction experienced by program participants
- staff qualifications to implement the program
- competing effects of factors outside the program (i.e., exogenous influences), and
- the degree of coordination among different program components and activities.

All of these factors can be used as explanatory factors in data analysis and as such provide a means of accounting for the effects of various events on the outcome of an intervention program. Variations in dosage, implementation, and contextual factors (e.g., school, student, and staff characteristics) may mediate or moderate the outcomes of an intervention. Including these factors in a causal model can assist in explaining the degree to which each factor contributed to or detracted from the achieved outcomes as well as explain why uniform outcomes were not observed across multiple intervention sites (Raizman et al., 1994).

Process data is not only useful in developing causal models of what, how and why certain intervention components are effective but also highlight to whom and under what circumstances an intervention (or its components) are most effective or ineffective. Conceptual models are necessary to facilitate an understanding of how the components of an intervention are expected to work in concert and to assure that vital elements of an intervention are measured (McGraw et al., 1994, 1996). Process information is important because it promotes standardized program implementation across sites and fine tuning of implementations when necessary in order to keep program execution consistent with its design and purpose (Rossi & Freeman, 1993).

Implementation data should be obtained not only on individuals receiving services but also on those delivering them. It is important to know the extent of training received by those involved in the actual program delivery because the amount of training received by these individuals may be linked to both the dosage of the intervention received by participants and the fidelity with which the program is implemented (McGraw et al., 1994). The level of coordination among program components and activities also is a good gauge to determining the quality of implementation (i.e., smoothness of program delivery) as well as the organizational context of program operations. This is a special concern for complex, multi-component school and community-based programs. As shown in Table 6.1, the interorganizational context for implementation can be measured through ratings

Table 6.1.   Perception of System Interorganizational Characteristics
(Listed below are statements that may apply to services for children
and families within this school/school district's geographic area.
Please select the response that best describes the situation.)

In this geographic area:	Always 4	Usually 3	Sometimes 2	Rarely 1	Never 0	Don't Know 8
Children and families receive adequate services as needed.						
Programs reach all children and families who need them.						
Services are easily accessible to families.						
Services for children and families are well-coordinated.						
Agencies share information and resources						
Referrals are shared between agencies.						
Interagency meetings occur.						
There is joint planning across agencies.						
There is problem-solving across agencies to fill gaps in services						
Agencies make children and families a very high priority in terms of providing services.						
Agencies are often in conflict with one another.						
Agencies have conflicting rules and eligibility requirements.						
Duplication of services is a problem.						
Children and families face barriers to obtaining services.						

by program staff including the extent to which information and resources are shared and whether joint planning activities occur. These and other data on implementation have been collected through the Service Delivery System Questionnaire (developed by the University of Wisconsin Center for Health Policy and Program Evaluation) in evaluations of the Comprehensive Child Development Program and the Head Start-Public School Transition Program.

It is only by documenting the implementation of a program that evaluators know what happened. This is of critical importance because an intervention's implementation is vulnerable to alterations at all levels of program delivery. In order to know whether the implementation is consistent with the program's design, evaluators must have a full and accurate accounting of this process (Scheirer, 1994; Rossi & Freeman, 1993). Implementation data should be collected not only from multiple sources but also through multiple means. The implementation data collected for a school-based intervention, might include weekly checklists which document the training of those responsible for the intervention, forms documenting both the amount and extent of program implementation by teachers and other school personnel, ongoing observations by evaluation staff that provide information which complements the data obtained through more structured instruments (Edmundson et al., 1994), school record data (e.g., grades, standardized test scores, disciplinary infractions, transcripts, etc.), campus profiles developed by school districts, telephone interviews with parents/guardians (Scheirer, 1994; Rossi & Freeman, 1993), and management information systems (Holden & Reynolds, 1997).

Implementation data collected by evaluators should document the number of visits (scheduled and unscheduled) made to monitor the implementation of a program and address questions and concerns of program staff. Monitoring of this type facilitates not only making changes in procedures or materials when necessary but also documenting such changes. The documentation of changes in an intervention's implementation provides additional data to include in subsequent analyses (McGraw et al., 1994). Comprehensive process evaluation should also include data on satisfaction and level of service provision from teachers, all school staff involved in some aspect of the intervention, parents, and personnel in the control schools (e.g., to determine whether control schools engaged in competing activities or programs) (Elder et al. 1994; Johnson et al., 1994; Raizman et al., 1994).

The use of program templates or profiles is one recent development to help evaluators assess how well different program components have worked and how well they reflect "best" practices. As Loucks-Horsley (1996) indicated, a program template is a "tool for standardizing the description of a program and facilitating comparison of the program's components with what research and practice say are effective" (p. 6). A template is organized to include a description of (a) best practices or program standards as determined from research, (b) the intended program components and activities, and finally (c) the actual program operations and activities as experienced by participants. See Scheirer (1996) for further details and examples.

EXAMPLES OF EVALUATIONS OF EDUCATIONAL PROGRAMS

Implementation evaluation is often neglected in the design or modification of a program. This can have dire consequences for a program's longevity particularly when it is necessary to provide accountability data. When a program's directors are unable to demonstrate that their program benefits those it serves, they face the prospect of having their funding reduced or simply discontinued. When program evaluation is included in a program's design, planners can identify outcome measures and if necessary develop, pilot and validate scales to be used when documenting the outcomes of an intervention. What follows are examples of how two well known educational interventions—Head Start and the School Development Program—were implemented and evaluated and the role of implementation evaluation in this process.

Head Start

Project Head Start, the national child development preschool program, was authorized through the Economic Opportunity Act in October 1964. In February 1965, the Head Start Planning Committee identified several objectives for what was the first federal program to coordinate health, parent involvement, social services, and education for poor pre-school age children in the United States. Among the objectives outlined for Head Start were to improve children's physical health, "establish patterns and expectations of success for children which will create a climate of confidence for their future learning efforts," increase parental involvement in children's education, and "improve the conceptual and verbal skills" of children (Zigler & Muenchow, 1992).

Head Start did not begin as a small pilot program but as a nationwide eight week summer program for 100,000 economically disadvantaged children. The program was up and running in 12 weeks. In some circles, it was called "project rush-rush" because of the speed with which it was designed and implemented. In its first two years, Head Start was a summer program, but in fiscal year 1966 its budget tripled and it became a year round program serving 500,000 children.

The creators of Head Start anticipated the program's evaluation and stated that "research and evaluation (of Head Start) should be part of local and national efforts arranged by (the) Office of Economic Opportunity through independent assessment of local programs to identity successful techniques and programs." However, as Zigler and Muenchow (1992) indicated, the 1965 evaluation of Head Start's first summer program was poorly planned and implemented, and the measures used to investigate process and outcome data were not validated. These and other problems (e.g., lack of appropriate resources) led to the decision to not evaluate the effects of the initial implementation.

In March 1968, the Office of Economic Opportunity issued a request for proposals for an overall evaluation of Head Start. The contract, awarded to Westinghouse Learning Corporation in conjunction with Ohio University, required that the evaluation be competed by April 1969. The Westinghouse/Ohio report (Cicirelli et al., 1969) was the first large-scale national study to evaluate the effects of Head Start participation on later school achievement. It has been widely criticized since it was issued. Among the chief criticisms of the Westinghouse/Ohio evaluation were that: (1) the assessment was too soon; (2) the evaluation should not have been based on the first summer of Head Start's existence; (3) the multiple goals of the program were not equally emphasized in the evaluation; and (4) the effects of different curricula were not examined (Zigler & Muenchow, 1992; Washington & Bailey, 1995). Some of these criticisms reflect the hurried nature of the Head Start implementation and others reflect the equally hurried nature of the Westinghouse/Ohio evaluation of Head Start.

The Westinghouse/Ohio Head Start evaluation assessed outcomes of the program using a post-test only nonequivalent groups design and not the process involved in its implementation. As reported by the evaluators (Cicirelli et al., 1969), implementation evaluation "was not a basic question of the study, and the time schedule did not permit intensive investigation of the program." In addition, because of the lack of process information "it was impossible to know in detail the actual program that these children experienced." A process evaluation of a multi-site intervention of this magnitude would have required documenting the steps of the intervention's implementation at a representative sample of Head Start sites, but this was not done.

Head Start's implementation would have made a process evaluation difficult for several reasons: (1) the program model was not fully specified and there were no precise objectives or standards by which to measure its fidelity of implementation, (2) programs varied from site to site and no uniform data were collected thus limiting the capacity to make cross-site comparisons, and (3) the timing of the initial program implementation and the Westinghouse evaluation limited the ability of program managers and evaluators to pilot the program and the instruments needed to assess process and outcomes.

While the findings of the Westinghouse study were largely unfavorable—no effects of the summer program were detected and the immediate positive effects of the full-year program on cognitive development for most students faded substantially by first grade—a process evaluation would have helped substantially. Not only would a process evaluation have helped to avoid a type III error (i.e., evaluating an unimplemented or poorly implemented program) but would have helped to distinguish the extent to which the unexpected findings were due to theory failure, implementation failure, and/or limitations of the research design and measures. Critiques of the evaluation emphasized these latter two reasons.

Moreover, attention to process evaluation could have documented significant variations among programs and between program components. Reanalyses of the Westinghouse data shed light on this implementation issue. Using previously unanalyzed data on indicators of implementation quality from a subsample of centers (collected from interviews conducted with Head Start staff), Wu (1991) reported that the Head Start centers analyzed in the Westinghouse study varied widely on such indicators as program duration (3 to 8 months for full-year programs), teacher-child ratios (1:4 to 1:16), the number of supplies and types of equipment (15 to 42), and the number of places to which trips were taken (3 to 13). Since all of these factors are known predictors of program quality, overall analyses were likely to underestimate the effects of program participation for many centers. Indeed, in a latent-variable analysis, Wu (1991) found that teacher-child ratio, number of trips, amount of supplies/equipment, and the extent of effort given to ascribed program objectives were all significantly associated with children's cognitive ability in grades 1 or 2 (net of family SES). Thus analyses of program delivery would have added a more wholistic portrait of the program and strengthened considerably the interpretation of findings.

The example of Head Start is clearly unique and many of its problems and successes are due to the social and political climate in which it was developed, implemented, and evaluated. Based on our earlier discussion in this chapter, it would have been preferable to have conducted a process evaluation prior to the outcome evaluation or at the very least to have conducted the process evaluation in tandem with the outcome evaluation. Moreover, given the difficulties inherent in implementing new programs, Head Start may also have benefitted from beginning as a small pilot program and undergoing extensive formative evaluation. Without evidence on implementation, the first reaction to less-than-expected results is often to attribute failures to the program theory or design rather than to problems in implementation or research procedures. Of course, many later better-designed studies showed positive effects on children's development (McKey et al., 1985). As Zigler and Muenchow (1992) discussed, the findings of the Westinghouse study, ironically, probably would have led to the termination of a smaller-scale Head Start program.

Nevertheless, a smaller-scale program would have offered developers the opportunity to fully articulate the specific objectives of the program, track the process of program implementation in a representative sample of programs, alter or eliminate those aspects of the program that were not working, and reinforce those that worked well.

The School Development Program

At its inception, the Head Start program was implemented widely and emphasized outcome evaluation over implementation evaluation. A different approach was followed in a small-scaled model educational intervention which began in the late 1960s. The School Development Program (SDP) chronicled in James Comer's book, *School Power* (1993), was designed as a theory-driven educational intervention in two New Haven schools. Recognizing that research in schools is hampered by a shortage of theories comprehensive enough to be useful, Comer and his colleagues chose a human ecology model, focusing on the interactions between people and their environment (i.e., how people influence their environment and are in turn influenced by their environment). Developers of the SDP hypothesized that the school climate, academic achievement and social growth of students would be improved with the application of social and behavioral science principles to the educational program. They identified five program goals: (1) modify the school climate to facilitate learning; (2) make statistically significant improvements in academic achievement (i.e., standardized test scores in reading and mathematics); (3) increase students' academic and occupational aspirations by improving their motivation and sense of mastery; (4) develop decision making patterns which promote collaboration between parents and school staff; and (5) develop organizational relationships among the schools/school district and child development and mental health specialists.

With these goals in mind the program was organized and operated to promote cooperation in the identification of both problems and solutions which enabled participants (i.e., parents, teachers, and school administrators) to internalize the goals and methods of the program and feel invested in the outcomes. The SDP was originally piloted at two schools. Although the program developers made a conscious decision early on not to conduct a full-scale evaluation of the SDP they recognized the importance of documenting the dosage and fidelity of the implementation and determined that a "low profile" evaluation for internal use would be necessary to both inform the development of educational and extra curricular programs and facilitate dissemination of the program. Since there were no control group comparison schools, only within school comparisons were made. The first year, the SDP's implementation was detailed in an observer's diary. Early measures included consumer evaluations from school staff, parents, and some students, studies of academic achievement, and educational and clinical support services. During the first five years of the program evaluation findings were inconclusive. Later evaluations indicated that major program goals were achieved (Comer, 1993).

A path analysis documented the connection between good program implementation, positive school climate and improved academic performance. Comer (1993) noted that good program implementation and positive school climate needed to be in place two to five years before there would be noticeable changes in academic performance (see also Anson, Cook, Habib, Grady, Haynes, & Comer, 1991 for details). According to Comer, between 1969 and 1984 students demonstrated steady gains in mathematics and reading achievement. From 1982-1984 students in SDP schools experienced a greater decline in school suspension, days absent, and corporal punishment than those in non-SDP schools. Recent studies of SDP students indicate that they have more positive self-concepts than non-SDP students. These findings suggest that successful programs are not implemented quickly but must be refined several times before optimal implementation and expected outcomes are achieved. In contrast to Head Start, the evaluators of the School Development Program avoided conducting an outcome evaluation until the program was satisfactorily debugged. This philosophy is in keeping with Campbell's (1984) recommendation to evaluate only "proud" programs. Admittedly, given the inclination of policymakers and administrators to want data on program effectiveness immediately, this strategy may need to be moderated in the evaluation of a large-scale, government-funded program.

SPECIAL PROBLEMS FOR PROGRAM IMPLEMENTATION

These two examples raise three issues evaluators often face in conducting evaluations. The first concerns the appropriate balance between process and outcome evaluation during the early years of program implementation. Certainly, large-scale or complex programs like Head Start and the School Development Program pose special challenges in this regard. Although our earlier point that successful program delivery is a fundamental assumption of outcome evaluation suggests that process evaluation should be highlighted early in evaluations of new programs, in some circumstances this may not be possible or even desirable. In the case of Head Start, the political pressure to justify a new national program in the War on Poverty was overwhelming and an extensive implementation evaluation would have exceeded the financial resources available. Alternatively, the early emphasis on process evaluation in the School Development Program was probably the best course of action. The program was small in scale. It was a total transformation of the school culture and organization, and parents and school personnel distrusted the researchers' commitment to the schools. Consequently, an early outcome evaluation would have been quite problematic.

The second issue raised by the evaluation examples is evaluating nonuniform treatment delivery. Due to local interests, resources, and needs, many educational programs and reforms are implemented nonuniformily by design. As would be expected, the effects of such programs will vary by the extent and quality of program implementation. This certainly was the case in the Head Start Program as well as in replications of the School Development Program. The challenges of conducting evaluations of nonuniform treatments would be magnified for large-scale block grants such as Title I as well as district-wide school reforms such as those in Chicago and other large cities (see Niemic & Walberg, 1993). As illustrated throughout this chapter, implementation evaluation is crucial in such programs.

A final issue raised by these examples that has implications for process evaluation is to which particular program operation does an evaluator attribute success if findings show gains for the program group? Often, the evaluator is interested primarily in change at the molar level—the total program package—rather than the unique contribution of different components or experiences. Implementation evaluation provides more refined information and insight about all the components and activities of the program that were delivered so that the level and quality of implementation can be compared across these components. Thus, the specific sources of the program effectiveness can be better understood. Evaluations of Head Start in the 1970s and 1980s, for example, have demonstrated positive effects on children's school achievement. Data on implementation fidelity across health, education, social service, or family support components could be used to investigate if child outcomes vary as a function of the quality of implementation between components. A similar evaluation approach may be used for the School Development Program. Process evaluation provides unique opportunities for linking specific program elements with program outcomes.

IMPLICATIONS FOR EVALUATION

The outcomes of an intervention are inextricably linked to its implementation and must be documented. A poorly or inconsistently implemented program is unlikely to produce the intended outcomes. Documenting the elements of an intervention as they are actually experienced by the target group is the only effective means of determining fidelity to program principles and of ensuring a fundamental assumption of outcome evaluation. Conducting the implementation evaluations of programs in schools is challenging for several reasons: (1) ethical issues (i.e., informed consent and confidentiality) must be addressed with extreme care; (2) intervention activities and methods of data collection must be acceptable to participating schools; (3) school staff have little experience with and/or investment in research; and (4) although the use of process measures may pose minimal risk to participants, it requires continual data collection which some school administrators and personnel may find disruptive (Lytle et al., 1994). When designing an intervention, program developers must be sensitive to these issues and recognize that a great deal of diplomacy will be required to successfully implement intervention programs.

Although participants benefit from programs like Head Start and the School Development Program, the positive effects of these educational interventions were not documented until each had been successfully implemented for several years. Effective program implementation takes time. If program developers and funding agencies are convinced that a program's underlying theory is sound, then programs should be piloted for one to two years and outcomes of the intervention should be formally assessed soon thereafter. This will provide time to thoroughly document elements of the intervention (including unanticipated events), test and refine instruments developed to record the process, and address reservations of all participants such as school administrators, teachers, students, and parents.

Certainly, evaluations of program implementation require significant amounts of time and effort on the part of researchers as well as financial resources. As discussed in this chapter, the benefits of implementation evaluations are many, and it is programs and participants that have the most to gain from their frequent use.

REFERENCES

Anson, A. R., Cook, T. D., Habib, F., Grady, M. K., Haynes, N., & Comer, J. P. (1991). The Comer school development program: A theoretical analysis. *Urban Education, 26*, 56-82.

Blalock, A. B. (Ed.). (1990). *Evaluating social programs at the state and local level.* Kalamazoo, MI: W. E. Upjohn Institute.

Campbell, D. T. (1984). Can we be scientific in applied social research? In R. F. Conner, D. G. Altman, & C. Jackson (Eds.), *Evaluation studies review annual* (Vol. 9, pp. 26-48). Beverly Hills, CA: Sage.

Charters, W. W., & Jones, J. E. (1973). On the risk of appraising non-events in program evaluation. *Evaluation Researcher, 2*(11), 5-7.

Chen, H. T. (1990). *Theory-driven evaluations.* Newbury Park, CA: Sage.

Cicirelli, V. G. et al. (1969). The impact of Head Start: An evaluation of the effects of Head Start on children's cognitive and affective development. Athens, OH: Ohio University and Westinghouse Learning Corporation.

Comer, J. (1993). *School Power: Implications of an intervention project* (Rev. ed.). New York: Free Press.

Davidson, W. S., & Redner, R. (1988). The prevention of juvenile delinquency: Diversion from the juvenile justice system. In R. H. Price et al. (Eds.), *14 ounces of prevention: A casebook for practitioners* (pp. 123-137). Washington, DC: APA.

Doernberger, C., & Zigler, E. (1993). America's Title I/Chapter I programs: Why the promise has not been met. In E. Zigler & S. J. Styfco (Eds.), *Head Start and beyond: A national plan for extended childhood intervention* (pp. 73-95). New Haven, CT: Yale University Press.

Edmundson, E.W., Luton, S.C., McGraw, S.A., Kelder, S.H., Layman, A.K., Smyth, M.H., Bachman, K.J., Pedersen, S.A., & Stone, E.J. (1994). CATCH: Classroom process evaluation in a multicenter trial. *Health Education Quarterly* (Supplement 2), S27-S50.

Elder, J.P., McGraw, S.A., Stone, E.J., Reed, D.B., Harsha, D.W., Greene, T., & Wambsgans, K.C. (1994). CATCH: Process evaluation of environmental factors and programs. *Health Education Quarterly* (Supplement 2), S107-S127.

Holden, K.C., & Reynolds, A. (1997). Process evaluation of W-2: What it is, why it is useful, and how to do it? *Evaluation comprehensive welfare reforms: A conference* (pp. 139-156). Special Report No. 69. Madison, WI: Institute for Research on Poverty, University of Wisconsin-Madison.

Loucks-Horsley, S. (1996). The design of templates as tools for formative evalution. In M. A. Scheirer (Ed.) *A user's guide to program templates: A new tool for evaluating program content* (pp. 5-24). San Francisco: Jossey-Bass.

Johnson, C.C., Osganian, S.K., Budman, S.B., Lytle, L.A., Barrera, E.P., Bonura, S.R., Wu, M.C., & Nader, P.R. (1994). CATCH: Family process evaluation in a multicenter trial. *Health Education Quarterly* (Supplement 2), S91-S106.

Lytle, L.A., Davidann, B.Z., Bachman, K., Edmundson, E.W., Johnson, C.C., Reeds, J.N., Wambsgans, K.C., & Budman, S. (1994). CATCH: Challenges of conducting process evaluation in a multicenter trial. *Health Education Quarterly* (Supplement 2), S129-S141.

McGraw, S.A., Stone, E.J., Osganian, S.K., Elder, J.P., Perry, C.L., Johnson, C.C., Parcel, G.S., Webber, L.S., & Luepker, R.V. (1994). Design of process evaluation within the child and adolescent trial for cardiovascular health (CATCH). *Health Education Quarterly* (Supplement 2), S5-S26.

McGraw, S.A., Sellers, D.E., Stone, E.J., Bebchuk, J., Edmundson, E.W., Johnson, C.C., Bachman, K.J., & Luepker, R.V. (1996). Using process data to explain outcomes: An illustration from the child and adolescent trial for cardiovascular health (CATCH). *Evaluation Review, 20*(3), 291-312.

McKey, R. H., Condelli, L., Ganson, H., Barrett, B. J., McConkey, C., & Plantz, M. C. (1985). *The impact of Head Start on children, families, and communities* (DHHS Publication No. OHDS 85-31193). Washington, DC: U. S. Government Printing Office.

Niemiec, R. P., & Walberg, H. J. (Eds.). (1993). Evaluating Chicago school reform. *New directions for program evaluation* (No. 59). San Francisco: Jossey-Bass.

Raizman, D.J., Montgomery, D.H., Osganian, S.K., Ebzery, M.K., Evans, M.A., Niklas, T.A., Zive, M.M., Hann, B.J., Snyder; M.P., & Clesi, A.L. (1994). CATCH: Food service program process evaluation in a multicenter trial. *Health Education Quarterly* (Supplement 2), S51-S71.

Reynolds, A. J. (1991). The effects of an experiment-based physical science program on students' content knowledge and process skills. *Journal of Educational Research, 84*, 296-302.

Rossi, P.H. & Freeman, H.E. (1993). *Evaluation: A systematic approach*. Thousand Oaks, CA: Sage.

Scheirer, M. A. (Ed.). (1996). *A user's guide to program templates: A new tool for evaluating program content*. New Directions for Evaluation (No. 72). San Francisco: Jossey-Bass.

Scheirer, M.A. (1994). Designing and using process evaluation. In J.S. Wholey, H.P. Hatry, & K.E. Newcomer (Eds.) *Handbook of practical program evaluation*. San Francisco: Jossey-Bass Publishers.

Washington, V. & Bailey, U.J.O. (1995). *Project Head Start: Models and strategies for the twenty-first century*. New York: Garland Publishing, Inc.

Weiss, C. H. (1972). *Evaluation research*. Englewood Cliffs, NJ: Prentice-Hall.

Weiss, C. H. (1987). Evaluating social programs: What have we learned? *Society, 25*, 40-45.

Wholey, J. (1987). Evaluability assessment: Developing program theory. In L. Bickman (Ed.), *Using program theory in evaluation*. (No. 33, pp. 77-92). San Francisco: Jossey-Bass.

Wu, P. (1991). *Structural equation models in the analysis of data from a nonequivalent group design: A reanalysis of the Westinghouse Head Start evaluation*. Unpublish doctoral disseration, Lehigh University.

Zigler, E. & Muenchow, S. (1992). *Head Start: The inside story of America's most successful educational experiment*. New York: Basic Books.

Chapter 7

QUALITATIVE, INTERPRETATIVE EVALUATION

Jennifer C. Greene

"What is the overall effectiveness of this educational innovation?" "To what extent has this new program reduced the achievement gap between white and black students in U.S. schools?" "Do students in this new curriculum perform better or display stronger interest and motivation than students in the regular school curriculum?" "Do the benefits of this new educational program outweigh the costs?"

These are the kinds of questions about educational quality and ongoing educational change that are appropriately addressed with familiar and commonly used approaches to program evaluation, most prominently, social experiments or quasi-experiments and large-scale survey techniques.

"What is the meaningfulness of this new educational program for the participating teachers and students? How does it connect—or not—to the critical points of leverage in their everyday lives?" "What important contextual characteristics explain how this statewide educational reform is being differentially understood and implemented at the local level?" "In what ways does this new curriculum promote intended 'problem solving skills' among participating students? What 'problem solving' learning processes do students experience and how are these processes linked to essential features of the curriculum reform?" "What value stances are being promoted by this new educational program and how consonant are these with the

Advances in Educational Productivity, Volume 7, pages 135-154.
Copyright © 1998 by JAI Press Inc.
All rights of reproduction in any form reserved.
ISBN: 0-7623-0253-4

diverse values characteristic of participating sites?"

These, in contrast, are questions about educational quality and change that are appropriately addressed with more recently developed qualitative approaches to program evaluation, approaches that emphasize onsite interactions and gathering of information about people's educational experiences using their own words and stories.

THE EMERGENCE OF QUALITATIVE EVALUATION

Qualitative approaches to program evaluation emerged within the U.S. evaluation community in the 1970s (Stake, 1967, 1975). These approaches constituted a significant response to two interrelated sets of changes in intellectual thought and societal beliefs (Cook, 1985; Greene, 1994; Greene & McClintock, 1985).

First, within the philosophy of science, the scientific paradigm[1] that had guided most work in both the natural and the social sciences since the Enlightenment era of the eighteenth century came under critical attack. Notably and nearly universally critiqued was this paradigm's assumption of *objectivity*, of the possibility of obtaining scientific observations that are untainted by the theoretical and value predispositions of the observer. As now argued by philosophers of varying stripes (Bernstein, 1983; Phillips, 1990), it is not possible to stand outside one's own historical, professional, and personal location and make pure or objective scientific observations or knowledge claims. Rather, all such observations and knowledge claims are inevitably filtered through and thus imbued with the particular experiential and conceptual lenses of the observer. Also critiqued—but with considerably more and continuing controversy important especially to the social sciences (e.g., House, 1992)—were traditional science's assumptions that (a) the social phenomena we endeavor to know are real and can be explained, predicted, and controlled by universal laws and (b) concomitantly, the business of science is to discover and verify these laws. Challenges to these *realist* assumptions came primarily from philosophical traditions of idealism, phenomenology, interpretivism and constructivism (Schwandt, 1994). These challengers argued that what is important to know and understand about social phenomena is the localized meaning that people construct about these phenomena as they interact with and experience them in concert with others. It is this localized meaning, not external laws, that most importantly influences people's beliefs and guides their actions. Therefore, the business of social science—and the nature of important social scientific knowledge claims—concern the meaningfulness of people's experiences, which are expected to be as diverse as the contexts of human interaction.

These and other critical challenges to the traditional scientific paradigm strongly shook up established doctrine within U.S. social sciences, especially the applied social sciences including educational research and evaluation. Amidst the whirlwinds of ensuing debates and controversies came multiple opportunities for alternative paradigms—such as those guiding qualitative social inquiry—to emerge, gain a foothold, and begin to flourish within the arenas of educational and social program evaluation.

The second major set of changes that encouraged, even enabled, the emergence and growth of qualitative approaches to evaluation were changes in the belief systems of the larger U.S. society during the 1960s and 1970s. Thomas Cook (1985) described these changes as:

1. The decline in authority accorded standard social science theory following the perceived failures of the Great Society (see also House, 1993).
2. The decline in authority accorded political figures following the Vietnam War, the Watergate scandal, and other government debacles.
3. An increase in the cultural, value and political pluralism of society, as manifest, for example, in the civil rights and women's movements.

There was considerable consonance between these shifts in societal thinking and the emergence of alternative paradigms for understanding and doing social science.

CHARACTERIZING QUALITATIVE EVALUATION

Once alternative but now reasonably well accepted by most evaluators and evaluation users alike, "qualitative evaluation" actually embraces a variety of different forms of and approaches to assessing the merit and worth of educational and social programs. These include the individual case study (Mabry, this volume), the purposeful selection of extreme cases in the search for "best practices," the chronicling of life histories, and the analysis of text and discourse. What these various approaches generally share are the following core tenets and assumptions of the *interpretive* inquiry paradigm and tradition (see also Schwandt, 1994; J.K. Smith, 1989).

- Human beliefs and behaviors are importantly guided by the *meanings* individuals construct about the events and interactions in their lives, by how people interpret and make sense of their own situations. So, what is important to know about human activities, such as an educational intervention, is how it is experienced and understood as meaningful, or not, by individual participants in this activity or intervention.
- Values and valuing are an essential part of meaning and thus of social scientific knowing. There can be "no facts without values, and different values can actually lead to different facts" (Smith, 1989, p. 111). So, interpretive study intentionally reveals and incorporates the value dimensions of lived experience.

- This experience and its meaningfulness are specific to a given *context,* and even further, to individual members within that context. That is, teachers within a single school may experience a schoolwide reform differently, depending on their individual histories, values, pedagogies, and so forth. And these experiences will overlap with but also depart from those of teachers in another school, attributable to differences in the schools' climates, histories, cultures, and norms. *Diversity and multiplicity* are therefore expected in the understandings attained in a qualitative evaluation.
- Qualitative evaluation results are also expected to convey in some depth the *complexities and idiosyncrasies* of human interactions and interpretations, while also weaving these complexities into a whole. This whole constitutes a *holistic portrayal* of the context studied, a portrayal marked by coherence, integrity, and credibility. That is, individual complexities do not have meaning in isolation from the whole of their context nor can the whole context be understood stripped of its complexities, tensions and fissures. [2]

So, qualitative evaluations are characteristically indepth studies of small samples, using mostly but not exclusively qualitative methods, such as interviews and observations, and focused on the understanding of *insiders'* perspectives and meanings. Indepth inquiry is required to develop credible understandings of insiders' experiences of, say, an educational reform in a given school district. Practical and resource constraints generally preclude the possibility of developing these credible understandings of inside meaning for more than a few insiders. The qualitative evaluator is thus challenged to weave the insider views that are constructed into a coherent and credible story of reform of relevance to the whole district. Qualitative methods—methods that use words and other non-numeric symbols (photographs, artifacts) to represent peoples' experiences—are generally preferred because they can best capture constructed and complex meanings. Similarly, qualitative evaluation results are characteristically conveyed through the reporting rubric of the story or the narrative, because a story can best portray the dynamic interplay of part and whole, the complexities of individual experiences and perspectives within the fabric of the whole.

To illustrate these fundamental tenets of qualitative evaluation, envision an evaluation of a high school curriculum reform in science, a reform which endeavors to help students develop higher-order and critical thinking skills, as well as scientific literacy and reasoning, rather than master multiple discrete facts. A major feature of the new curriculum is that it is structured around everyday, real world problems, like recombinant DNA engineering, oil spills in the ocean, and coastal erosion. Another major feature of the new program is its extensive use of student work groups.

To address evaluative questions about the quality and effectiveness of this science curriculum reform, the qualitative evaluator would seek primarily to understand how the program is experienced by individual students and teachers in particular classroom contexts. For it is in these experiences that the meanings of program quality and effectiveness, and thereby the meanings of educational development and productivity, are shaped and molded. The qualitative evaluator of this science reform would likely sample for data gathering classrooms, teachers, and students to obtain heterogeneous samples on relevant characteristics (for example, demographics, scientific career aspirations, teaching philosophy), in order to capture the full range of diversity of experiences and meanings expected. The evaluator would likely sample two or three classrooms for extended observation, in order to develop a descriptive portrayal of what actually happens in the new science classes. How do teachers use real world contexts to engage students in the learning of scientific concepts and reasoning? What is a typical lesson like? How do the student work groups function? The evaluator would likely sample 10-15 students for individual interviews and reviews of their science portfolios, in order to understand what students find meaningful in their science classes and how this is reflected in their work. The evaluator would also likely sample some teachers, administrators, and possibly parents or community members for interviews regarding their perceptions of and experiences with the new science program. What are the range of views regarding the practical and group learning emphases of the new program? What agreements and disagreements exist within the community about the kinds of learning and science achievement being fostered in the new program? Alternatively, parent/community perceptions might be gathered with a mail survey. Student learning and achievement data, based on a wide variety of indicators, would be extracted from school records. All of these data would be analyzed for major themes and categories of meaning, which would then be rewoven into a holistic portrayal about program quality and effectiveness as grounded in insider experiences and meanings. This portrayal would be offered as a representation of those studied, with possible but not assured extensions and relevance to those not studied. This portrayal would be filled with snapshots of science learning, vignettes about classroom rhythms and dynamics, stories of individual accomplishment and perhaps disengagement, and other ways of connecting this educational intervention with individual lives. This portrayal would thus offer a contextualized, complex, dynamic, value-laden, and necessarily partial statement about educational productivity. This is what qualitative evaluation does.

EVALUATING QUALITATIVE EVALUATION

What has qualitative evaluation contributed to the development and refinement of defensible educational policies and effective educational programs? In what ways has qualitative evaluation served to strengthen and enhance these policies and programs? And in counterpoint, what have been the shortcomings of qualitative evaluation? In what ways has it not fulfilled its promises or not operated to improve our nation's educational policies and programs? These meta-evaluative questions are addressed in this section, which is structured around extended examples from the field.

To preview this discussion, I will argue that qualitative evaluation has served most importantly to disaggregate summary statistics into individual lives and to provide insight into what an educational policy or program looks like and means to those whose lives it is supposed to affect. For example, through Public Law 94-142 passed in 1984, thousands of children with disabilities have been "mainstreamed" into "regular" classrooms for part or all of the school day. Summary statistics offer counts of how many and what kinds of children this policy has affected. They also tell us whether or not the *average* disabled child achieves better, feels more efficacious, or perceives a reduced sense of difference from her peers in her mainstreamed classroom compared to her special education classroom. These statistics represent important information. But, also important are the different shapes, hues, and facets that give distinctive form and meaning to individual children's mainstreaming experiences—the cruelty of a deaf child's in-class interpreter, the frustration of an autistic child's regular teacher, the wall of peer indifference that keeps an emotionally disturbed child isolated, as well as a retarded child's pride in completing a math lesson and his wheelchair-bound classmate's thrill at playing basketball in the gym with all the other children. Qualitative evaluation has given substance, contextual relevance, and meaning to summary statistics, favoring not the average experience but the full richness of human experience in all of its diversity and complexity.

Qualitative evaluation has additionally helped to advance a broadening of evaluation's purview from an exclusive emphasis on program outcomes to the routine inclusion of program processes. Significant attention to program processes, to the meaningfulness of program experiences in context, is the very essence of qualitative, interpretive evaluation. Rather than view those being studied only as "windows onto universal psychological laws or indicators of treatment effects" (Graue & Walsh, 1995, p. 136), qualitative evaluators have encouraged others to focus instead on the meaningfulness of human action in its dynamic, relational, historical, and cultural context (Graue & Walsh, 1995). Comprehensive evaluations now regularly provide insight in the soundness of the program design, the quality of its implementation, and the variability of both site-to-site and participant-to-participant experiences in the program, in addition to the nature and extent of program outcomes. This substantially broader reach of evaluation generates greater program understanding and more explanatory power, specifically about *why and how* certain outcomes were attained or not. It also makes evaluation more useful to audiences other than policymakers, including, for example, program planners, program administrators and staff, and program participants.

Qualitative evaluation has also importantly contributed to an examination of the value claims and assumptions that accompany educational policies and programs. Just as interpretivism intentionally surfaces the value dimensions of experience, qualitative evaluations reveal for critical examination the oft-hidden premises and stances of educational interventions. What kind of student learning, and in what domains, is most important as we approach the twenty-first century? In a given educational reform, what are the differential values accorded excellence and equity of student learning? What assumptions about students' abilities to learn are represented in an innovative science curriculum or a technology-driven interdisciplinary reform? Part and parcel of qualitative evaluation is an engagement with these kinds of issues.

In these ways—by grounding program evaluation and therefore judgments about program effectiveness in participant experiences and meanings and in the values accompanying these varied experiences and meanings—qualitative evaluation has significantly *enlightened* (Weiss, 1977) public understanding and debate about important educational issues.

A reflective critique of the ways in which qualitative evaluation has not well served educational policymakers, program planners, and practitioners yields two sets of considerations. First is a set of considerations that are importantly constitutive of qualitative evaluation but require some thoughtful tempering and moderation in implementation rather than rigid adherence. These include the context-specificity and thus lack of generality of qualitative evaluation findings and, relatedly, their complexity. Qualitative evaluators have importantly helped to educate decision makers about the idiosyncratic, deep and inherent complexities of human phenomena—challenges and accomplishments alike. But, offering too much complexity can immobilize those charged with making decisions, and, at times, qualitative evaluators have done just that. Second, qualitative evaluators can be criticized for abdicating too much traditional scientific responsibility and for therefore having less impact than they could (and should) on our nation's educational system. In particular, by rejecting objectivity in favor of celebrating subjective insights and knowledge claims, and by discounting the relevance of existing theory and past research in favor of a "grounded and emic"[3] understanding of a particular context, qualitative inquirers have become good storytellers. Good stories illuminate the human condition, but don't usually offer specific solutions or recommend alternative endings, each based on different value stances and perspectives. It is time for qualitative evaluators to do more than tell good stories; it is time for them to reclaim their full responsibilities as scientific citizens (Greene, 1994, 1996).

These contributions and limitations of qualitative educational evaluation will be discussed and illustrated through a critique of three selected case examples.

Case Example #1: School Improvement, Teacher Professional Development, and Chicago School Reform

The first case example comes from the multi-year study of school improvement and teacher professional development in the Chicago city schools conducted by Robert Stake and colleagues at the Center for Instructional Research and Curriculum Evaluation (CIRCE) for the Chicago-based Teachers Academy for Mathematics and Science. Robert Stake's original work in responsive and qualitative evaluation (Stake, 1967, 1975, 1995) has importantly influenced the present generation of qualitative evaluators. The example presented herein is characteristic of Stake's approach and thus of much qualitative evaluation practice today.

The example is a case study of one Chicago K-8 school during the 1995-1996 school year, specifically, a study of the interconnections of the teachers' participation in professional development opportunities in math and science with the school's improvement and restructuring initiatives (Whiteaker, 1996). This study took place during the eighth year of the Chicago public school system's reform initiative, which emphasized decentralization of significant educational authority and control from a central administration to each local school. "It was theorized that local control would allow schools the autonomy to make the changes needed within the unique settings of their communities to improve the quality of education for their students" (Whiteaker, 1996, p. 89). The school studied in this evaluation, the Wilson School, was in its second year of participation in the professional development program of the Teachers Academy for Mathematics and Science.

Locating Evaluative Understanding in Context

As noted, the sponsor and client of this evaluation was the Teachers Academy for Mathematics and Science. Yet, in this case study the evaluator did not endeavor to isolate teachers' participation in this Academy and assess its specific, unique effects on school change or student achievement. This evaluation approach would decontextualize and thus fracture the meaning of these teachers' professional development. Rather, befitting the assumptions of a qualitative approach, Mikka Whiteaker sought to *holistically* describe and understand teachers' participation in the Academy as it connected to the overall rhythms of school life and to the evolving dynamics of school change in this particular school.

> [This evaluation sought to] describe the process of change at Wilson as the staff worked to improve mathematics and science learning. These improvement efforts were undertaken partly

through participation in the Teachers Academy.... Woven throughout the [evaluation] discussion and then articulated more explicitly towards the end [were] the different roles the Teachers Academy played in the process of school improvement at this particular school (Whiteaker, 1996, pp. 89, 98).

An assessment of the contributions of the Teachers Academy to Wilson School change and improvement was thus conducted *holistically and contextually.* This is a significant defining characteristic of qualitative evaluation.

The evaluation report for this case study thus appropriately offers at the outset critical contextual information. This includes a brief history of the school's improvement efforts, a vivid description of school classrooms and corridors, and descriptions of the major cast of characters—the student body, the teachers and teacher leaders, and the administrators. This descriptive portrayal—developed from considerable formal and informal observation on site—suits the narrative framework adopted for much qualitative evaluation reporting. More substantively, this contextual description is essential grounding for the interpretive accounts of teacher development and school improvement that follow. Again, these accounts are meaningful only in context. For illustrative purposes, snippets from the contextual description Whiteaker provided in her case report follow.

School pride. Wilson School was located in a south Chicago neighborhood teetering on the edge. With many factories in the area shut down, unemployment and frustration ran high. Temptation was great to give in to the poverty, crime and violence that plagued so many other communities in Chicago. Give in, they have not. At Wilson everyone was working hard to build a strong sense of community with their students and instill within them a sense of pride in themselves and their academic achievements. The front cover of Wilson's school improvement plan read, "Wilson pride is Wilson's power" (pp. 91-92).

Every day at Wilson began via public address. From the loud speakers in every room came the voices of students reciting the Pledge of Allegiance in both English and Spanish. Then a third student stepped to the microphone and we heard the Wilson Motto, further expounding the necessity of pride and community. "I am a student of the Wilson School community. I am committed to developing self esteem, respect for myself and others, a desire to learn, and the ability to make wise decisions. I will help to maintain a safe environment conducive to learning. I will be informed, optimistic, respectful, and open-minded. I will be a good citizen. I believe all children can learn and I will learn all I can and receive the best possible education. I will work to reach my full potential" (p. 92).

Wilson students. Wilson students were clearly at the center of all improvement efforts at the school.... 60% of the students at Wilson were African-Americans and 34% Latino/a. Students from low-income families made up 77% of the student population and 23% of students were "Limited-English proficient."... The attendance rate was 93%, with less than one percent displaying chronic truancy. This reflected Wilson student commitment to participating in the work of the school (p. 93).

Wilson teachers. The teachers of Wilson took pride in their school and the work they did. Everyone was well-dressed, greeting each other and students warmly in the halls. Raised voices were not frequently heard (p. 92).

School environment. Student work hung on the walls of classrooms along with posters of celebrities espousing the importance of education and hard work. Even the hallways boasted student art work amid the displays of information about different ethnic holidays and customs (p. 92).

Wilson leadership. [The] Principal and Assistant Principal enjoyed the high regard of the staff and teachers at Wilson.... The Principal, played [the role of] supporter and "instructional leader." According to her, principals should support the teachers' efforts to improve, first through the allocation of funds and time to participate.... Teachers frequently remarked that they felt they had [the Principal's] support for their participation in the Teacher Academy activities. [She] did not only support the teachers from the sidelines, but also attended Teachers Academy meetings, especially those for principals.... [She also] actively participated on Wilson's Leadership Team (pp. 93-94).

Centering Evaluative Understanding in Program Experiences

In addition to this grounding in context, Whiteaker's case study evaluation of Wilson School well characterizes the significant attention to program processes and interactions, rather than only program outcomes, that constitutes an important contribution of qualitative evaluation to the field. Information on program outcomes alone does not convey the ways in which the program was meaningful, or not, for participants. Nor does outcome information by itself inform our understanding of how and why a given program is effective, or not. In this case evaluation of Wilson School, the process of teacher development and the interconnected process of school improvement are both accorded significant attention, as illustrated next.

Whiteaker locates the teachers' own professional development in science and mathematics instruction *within* the school's living philosophy of school improvement. This philosophy was one of continuous growth, involving an ongoing cyclical process of reflection and adjustment and—during the school's second year of work with the Teachers Academy—"moving from a more individualistic view to an understanding of how individuals are part of broader change within the school... from an individualistic view of change toward one that is systemic, broader, and more coherent" (pp. 100-101). While the school principal provided the vision and the concrete support for this philosophy of school improvement, "much of the burden of school change rested on the shoulders of the teachers. It was the teachers who were being asked, not only to change the way they taught mathematics and science, but to make a fundamental shift in the way they thought about education and teaching in general" (p. 102).

At the core of this shift involved teacher understanding of a new constructivist pedagogy, which was integral to the Teachers Academy workshops on science and math instruction. Observations and interviews revealed that some Wilson teachers were learning and implementing in their classrooms some parts of this constructivist pedagogy, specifically manipulatives and hands-on activities, but were rarely engaging their students in reflective discussions on the math or science concepts underlying these activities. Other teachers conveyed greater comfort with an active-reflective teaching style and with students as legitimate contributors of knowledge in the classroom. The extent to which Wilson teachers had internalized constructivist pedagogy was thus variable and somewhat unclear. Whiteaker hypothesized that some teachers were in the process of grappling with concrete aspects of constructivism, like the hands-on activities, en route to developing a full, integrated intellectual understanding of constructivism. That is, paralleling the movement in the overall school change process from the micro level of the individual to the broader level of the system, teacher internalization of constructivist pedagogy reflected this same movement from individual action to broader philosophy, from part to whole. In sum,

> there was a movement within Wilson to move mathematics and science professional development away from an individual endeavor, to a cohesive vision of change at the school level. Teachers were beginning to see their improvement efforts as part of a larger improvement effort throughout the school. They were working collaboratively with other teachers to make structural changes in the school day [such as double periods for labs] to accommodate their new visions of mathematics and science teaching (p. 109).

Grounding Evaluative Interpretations in the Meanings of Lived Experience

This portrayal of the dynamic, intertwined *processes* of teacher development and school improvement affords considerable insight into ways in which change at Wilson School was being experienced by teachers. These experiences then constitute the essential forum for probing and analyzing the meaningfulness and effectiveness of these change endeavors. Whiteaker offers such probes in two domains: (1) the roles played by the Teachers Academy and (2) the linkages between Wilson's school improvement endeavors and what has been learned about successful teacher development and educational reform elsewhere.

The Teachers Academy was viewed as providing essential expertise and resources to the Wilson teachers—in math and science teaching, as well as in technology and school improvement—within a relationship marked by responsiveness and collaboration. Wilson teachers set the agenda for this relationship and Academy staff provided requested resources and guidance, as well as forums "for the exchange and discussion of information coming from the local, state and even national levels" (p. 117). Beyond this relationship, the Teachers Academy was not as helpful in directly supporting connections between teachers' professional development in math and science and overall school improvement, connections that Wilson teachers were therefore struggling with on their own. This analysis of the specific contributions of the Teachers Academy to Wilson School's "continuous growth" offers interpretive meaning that is grounded in the lived experiences of Wilson School's educational community.

Going Beyond the Story

With respect to the larger context of educational reform, Whiteaker, like most qualitative evaluators, claimed that "understanding the process of change at one school is ground for insight into where we can focus our efforts as researchers, educators and professional development providers to facilitate change in education" (p. 89). This claim is highly consonant with the advice of master evaluator Lee Cronbach (1980, 1982), whose work has significantly shaped the contemporary landscape of educational evaluation. Although not an active proponent of qualitative methods, Cronbach has vigorously advocated for indepth studies of educational interventions in individual sites in order to gain meaningful insight into the nature of the educational problems and challenges being addressed *and* into the success of the particular intervention being evaluated. "The better and more widely the workings of social programs are understood, the more rapidly policy will evolve and the more the programs will contribute to a better quality of life" (Cronbach, 1980, p. 3). The accumulation of deep and meaningful insights across settings, argued Cronbach, will do more to advance our understanding of educational problems than large, complex "horse race" studies. These studies focus only on identifying the best or "winning" intervention based on necessarily limited outcome indicators of teacher development, school change, or student achievement.

But where does the responsibility for accumulating such insights and acting upon them lie? Specifically, what is the evaluator's public accountability and responsibility for the political and action consequences of her or his work? Too many qualitative evaluators have highlighted the uniqueness of their settings and their stories and underplayed the commonalities, broader lessons learned, and significant action implications. Too many qualitative evaluators have contributed only their own stories and eschewed responsibility for how these stories are read, understood, and used. Whiteaker's case study, as well as the larger Teachers Academy evaluation of which it was a part, took important steps in the direction of connecting individual stories to broader learnings and actions. Several research-based models of teacher development and school reform contributed to both the framing and the analysis of the Wilson School and other case studies. But, like most qualitative evaluations, Whiteaker's case study findings and implications did not venture very far from the specific context of Wilson School.

Decision makers, planners, and practitioners are thus given a rich, grounded, complex story of the roles played by the Teachers Academy interventions in the intertwined processes of teacher development and school improvement in one school during one school year. The story honors the diversity and complexity of human experience. It relates the process of change, not just the outcomes. It anchors significance in the lived experiences of members of the setting being studied. But, the story doesn't automatically inform the larger conversation about educational reform, nor automatically guide decisions about promising directions for reform. In these ways, qualitative evaluation underfulfills its potential and does not fully meet its responsibilities. To do so will require qualitative evaluators to step across the cracks that separate individual contexts, to find the simple nugget of transferable insight amidst the complexity of the narrative, and to intentionally connect the themes of this one story to outside and broader concerns and issues. Qualitative evaluators need to offer more than their individual stories.

Case Example #2: But Stories are so Powerful...
Head Start Pride Countered by Desegregation

The Head Start program in "Centerville" was well integrated into the culture and rhythms of life in this predominantly African-American, poor, rural town in North Carolina. A qualitative case study of this program in its cultural context (Philipsen & Agnew, 1995) revealed the following.

> [The] mothers and teachers of young children in Centerville are deeply connected to and involved with their [Head Start] school, a school located in the community and run by teachers who share the children's racial background.... [So] the black students... are exposed daily to black role models.

> Besides the deep sense of belonging to the school shared by children, parents, and teachers alike, the Head Start curriculum is recognized as relevant to children's lives.... [possessing] a sense of cultural congruence. The Head Start curriculum—what was being taught and how it was being taught—coincided with and reinforced values taught at home.

> Another striking feature characterizes the meaning of Head Start for those involved: the culture of Head Start is constructed as a collective orientation, an orientation within which the needs of the group are stressed and communal values rather than individuality are emphasized.

> Said one grandmother, "Ya know, Head Start works with the whole family, not just the children but also the mothers and fathers and sometimes even the grandparents like me. I take care of the kids when their mother works and get to visit with the Head Start teacher who comes to our house."

> [Overall] this Head Start program... [provided] a positive experience where an initial sense of cultural identity and strength is tied to positive attitudes toward school (Philipsen & Agnew, 1995, pp. 51-53).

Centerville's Head Start program has been a part of the community since 1967. In 1969 the white public school in Centerville was closed as part of a countywide consolidation effort. In 1972 the black school was closed for desegregation objectives. Since then the older black students in Centerville have been bused to a number of predominantly white schools in three neighboring communities. African-American students are regularly reassigned from one school to another to meet desired racial mixes.

How well and meaningfully do these older students display the "cultural identity and pride" and "positive attitudes toward school" found for Centerville's Head Start graduates?

The [older] students from Centerville feel deeply alienated from the schools they attend. They do not have a sense of belonging to the predominantly white communities in which the schools are located (p. 53).

[For these students] schools do not provide the space to talk about "things that really matter."... Teachers are perceived to be non-supportive, particularly for those students who need special help (p. 54).

[Lost with the closing of the town's] all-black school is a sense among Centerville's parents that teachers care for the young. The culturally sanctioned ways of disciplining children in school and conducting the daily business of teaching were fundamentally different from what [these parents] perceived as the non-caring and liberal pedagogy of today's teachers in faraway schools.... [According to black parents, these teachers] have lost their moral authority [and] are not truly caring because they are not in control... they do not regulate, oversee, and discipline to the extent necessary to make sure that students learn, behave, and ultimately succeed...."out there" in a basically hostile and racist environment (pp. 54-56).

Said one highly frustrated parent, "Desegregation turned out to be a one-way street. Only the blacks have taken over the white values, and no whites took over black values" (p. 54).

The inquirers analyze this disjuncture between Head Start and later schooling for Centerville's African-American children as a rupture of schooling from the cultural fabric of its community context. "The positive effects of Head Start are in danger once the powerful triad of community, school, and students is broken" (p. 59).

[More fundamentally] by losing its schools, the black community [in Centerville has] lost education as a vehicle for social mobility. They have lost an essential tool in their struggle to keep up in an ever-changing and highly competitive world.... Except for the Head Start program, the black community in Centerville no longer has control over the one institution that carried the main burden of preparing the young for their future in a hostile white environment (p. 60).

Yes indeed, qualitative inquirers can tell powerful stories, as well illustrated by this story of culturally disrupted schooling in one poor, rural African-American community. The story is powerful in considerable measure because it *is* qualitative, that is, concerned with the meaningfulness and diversity of lived experience, thickly embedded in a cultural context, dynamic, and poignantly told largely from the eyes and in the words of the "real" people who live in Centerville. The story is powerful also because it surfaces some of the value dimensions of the larger societal and social policy issues to which it is connected through its explicit stance of support for the importance of "cultural identity and strength" for African-American children. The inquirers maintained this support while also acknowledging that Centerville's Head Start graduates were underprepared for the academic demands of public schooling. As noted, this attention to the sometimes discordant value dimensions of human experience is a significant asset of qualitative inquiry.

And further, it is precisely because qualitative stories can be so powerful that qualitative evaluators must more directly engage with these larger societal and policy issues. For this Centerville story, these issues include: Which domains of education are appropriately locally controlled? What are other important value dimensions of this policy conflict between Head Start and desegregation? And what action alternatives exist for resource-limited small rural communities? The third and final case example offers a more extensive illustration of the ways in which qualitative evaluators can offer more than their stories, powerful as they may be.

Case Example #3: The Constructed Meanings of "At Risk" in State-level Policymaking

In a qualitative policy evaluation that incorporated linguistic and cultural analyses, Margaret Placier (1993) traced the meaning of the term "at risk" in the educational policy making context of one state, Arizona. Placier personally interviewed and also analyzed the taped speeches of key state-level policy actors, as well as examined the minutes of state legislative education committees and the State Board of Education to trace the evolution of historical events. From a qualitative analysis of these data, Placier argued that Arizona's at risk educational policymaking reflected traditionalist assumptions about poor minority children and reinforced policy habits of strong state control, efficiency, and accountability. The policy process in Arizona, that is, did not embrace the reformist potential of the at risk construct, and, as a result, some resources were squandered and some children continued to suffer unnecessarily. What follows is an elaboration of Placier's argument, along with its implications for augmenting qualitative evaluation's authority and clout.

Understanding the Context

First, Placier presented her findings as integrally connected to the overall context of educational policy making in Arizona in the late 1980s and early 1990s. Using previous research, she characterized this context as "traditionalistic," meaning that economic elites (in Arizona, agriculture and mining) dominated educational policymaking, casting it largely as a means for economic development. Relatedly, the dominant value in Arizona educational policymaking was efficiency, associated with such policy mechanisms as accountability. The state Legislature was the most powerful policymaking body. The superintendent of public instruction had recently worked hard to strengthen the professional expertise and status of the education department through, for example, better information systems and statistical reporting. These efforts served in large part to "reinforce the traditionalistic tightfistedness of the Legislature with scientific rationality" (Placier, 1993, p. 384).

Within this context, Placier next observed that there was bipartisan consensus that at-risk students were a serious problem for the state's economy and that an early intervention model, targeted toward certain students, was the most effective solution. As discussed below, Placier then analyzed the roles these consensual assumptions played in Arizona's at-risk policy making, with particular attention to the meanings ascribed to the term and concept of "at risk."

Explaining the Policy Consensus

Said a Senate education analyst,

> When *at risk* arrived on the national scene, it seemed to fit readily into [ongoing state] discussions as a label for students legislators were *already* interested in targeting: young children from poor homes, with presumed language deficits (p. 385, emphasis in original).

> [This] exemplifies an epidemiological approach to student failure, a belief that backgrounds of certain students (risk factors) make them susceptible to school failure and criminality (p. 385).

> [The term] *at risk* was a replacement [then] for previous descriptors for the same stigmatized group: poor, minority students...[but was a more] politically useful [and] desirable term because it was neutral with regard to race and class; it allowed policymakers to avoid ethnicity...(p. 386).

This was one factor in the *at risk* policy consensus in Arizona. Another significant factor promoting consensus came from the business sector, where concerns were expressed about future shortages of skilled and professional labor, thus supporting the idea that "education in Arizona was often defined as an economic function" (p. 387).

Understanding the Framing of At Risk Policies and Programs

> Arizona's first official at-risk legislation originated early in 1988 as SB 1328, "Children at Risk Prevention Programs," a bill proposing 4-year pilot projects for grades K-3.... [That is] Arizona policymakers adopted competitive, discretionary grants as their response to at-risk students [even though] this option was the least preferred among [a 1990 national survey of] state policymakers and national experts....Although this option rates low in effectiveness at addressing the problem, it achieves the goals of "state level management and control," [including] efficiency [and] accountability... the goals most congruent with Arizona's state policy culture (pp. 387-388).

Analyzing how At-Risk Grant Giving was Depoliticized Through Statistics

Arizona's at risk policy thus provided limited funds, to be allocated on a competitive basis. But, as said one professional staff member,

> When you have limited funds...you have to set up some sort of criteria for the people to compete. Otherwise, you're into a whole issue of favoritism or why would you give it to this district over that district? (p. 390)

Toward policy implementation, the education department recommended that program allocations be based on proposal quality *and* on program eligibility. They further proposed five specific eligibility criteria for K-3, which were "high student mobility, high absenteeism, high proportion of students scoring below the 25th percentile on the ITBS, high proportion of limited English proficiency students, and low socioeconomic status of students, [thus] proposing explicit meanings of *at risk*" (p. 389). The department chose these criteria primarily "because the state had comparable statistics on them from all districts" (p. 389), rather than because these criteria offered conceptually or empirically defensible meaning to the at risk construct. Further, for accountability purposes,

> the department imposed [again] an epidemiological model on districts, mandating that interventions be "targeted" at individual students identified as at risk. Program evaluations would be based on targeted students' academic progress, and continued receipt of funds would depend on positive results (p. 391).

Making Connections to Broader Issues and Value Stances

Placier's interpretation of Arizona's at risk policymaking includes the following critique. She argues that the education department's statistical criteria for at risk funding eligibility enabled policymakers to claim a fair, scientific, depoliticized way of distributing limited funds, "even though compliance with department expectations...and the efficiency norm were actually the major determinants....Thus, the definition and proposal-screening processes further consolidated state power in relation to local districts" (p. 392).

> Beyond this [there is] the question of how certain Arizona districts came to be defined as at risk, based on characteristics of their student populations and surrounding communities....The neutral term *at risk,*...did not invite the drawing of uncomfortable conclusions. [Yet] at-risk rankings were the [direct] result of historical conditions of inequitable funding, rural isolation from centers of power, racial and linguistic discrimination, and educational practices that have often been insensitive to local community needs (p. 392).

And thus Connecting Individual Stories to Broader Debates and to Action Recommendations

Placier's work offers an enlightening, contextually grounded story of educational policymaking in one state related to one group of children. Moreover, her work goes beyond this story in two ways. First, Placier links the story to broader critical concerns in contemporary educational discourse—concerns like equity and discrimination in educational opportunities, and tensions like personal versus economic development in educational outcomes. Second, Placier's work issues a challenge to Arizona's educational policy community—a challenge to think again about the meanings of being at risk and to reframe their policy conversation about at risk children around aspirations of racial, ethnic, and linguistic equity in the state, rather than maintenance of state control or policymaker comfort. With these explicit linkages to key issues within the broader educational policy arena and with this challenge to Arizona policy makers, Placier's work goes beyond an indepth understanding of the narrative to the generation of concrete recommendations for action and change, thereby involving and investing the inquirer in the consequences of her own inquiry.

Placier's work thus suggests some pathways for the qualitative evaluator to go beyond his/her story. These pathways require the evaluator to assert rather than subdue the value consciousness of qualitative inquiry, as Placier did in her critique of Arizona's policy history with at risk children. Such assertion implies an engagement with the value dimensions of human experience, an engagement that in turn implies responsibility for telling more than a good story, for linking the individual story to the larger conversation, the particular to its whole.

REPRISE

To be productive is to be effective in bringing about, to yield results or benefits of significance. Educational productivity is thus the effective yield of significant results, namely, the learning and socialization of children. What qualitative approaches to evaluating educational productivity most importantly do is to weave narrative tapestries that richly portray the contextual complexities and experiential meanings of what and how well children learn in schools. Results are connected to their generative processes and to their underlying value stances in qualitative evaluation. Through powerful stories, these complexities, processes, and values—in their full and rich human diversity—are revealed for engagement and discussion in qualitative evaluation. It is now time for qualitative evaluators to assume more active responsibility for the implications of their stories, and thereby to assert greater public accountability and enhanced policy authority.

This discussion suggests some of the ways qualitative evaluators can begin to fulfill their responsibilities as public scientists. They can make linkages between the drama of the particular context studied and the issues and concerns of the larger policy debates. They can make explicit the value dimensions of their own story and make connections between these dimensions and significant challenges in contemporary public life. They can offer specific ideas and directions for change and improvement. And they can assume some responsibility for participating in important action and change processes, not just critiquing or commenting on them from afar.

NOTES

1. From Thomas Kuhn, a social scientific paradigm is construed as "a set of interlocking philosophical assumptions and stances about knowledge, our social world, our ability to know that world,

and our reasons for knowing it—assumptions that collectively warrant certain methods, certain knowledge claims, and certain actions on those claims. A paradigm frames and guides a particular orientation to social inquiry, including what questions to ask, what methods to use, what knowledge claims to strive for, and what defines high-quality work" (Greene and Caracelli, 1997, p. 6).

2.   This interdependence of part and whole is derived from the *hermeneutic* tradition of interpret-. ing religious texts and, more recently, many other forms of symbolic meaning as well (J.K. Smith, 1989).

3.   Within the interpretive framework, "emic" refers to insider and "etic" refers to outsider understandings and meanings.

REFERENCES

Bernstein, R.J. (1983). *Beyond objectivism and relativism*. Philadelphia: University of Pennsylvania Press.

Cronbach, L.J. (1980). *Toward reform of program evaluation*. San Francisco: Jossey-Bass.

Cronbach, L.J. (1982). *Designing evaluations of educational and social programs*. San Francisco: Jossey-Bass.

Cook, T.D. (1985). Postpositivist critical multiplism. In L. Shotland & M.M. Mark (Eds.), *Social science and social policy* (pp. 21-62). Thousand Oaks, CA: Sage.

Graue, M.E., & Walsh, D.J. (1995). Children in context: Interpreting the here and now of children's lives. In J.A. Hatch (Ed.), *Qualitative research in early childhood settings* (pp. 135-154). Westport, CT: Praeger.

Greene, J.C. (1994). Qualitative program evaluation: Practice and promise. In N.K. Denzin & Y.S. Lincoln (Eds.), *Handbook of qualitative research* (pp. 530-544). Thousand Oaks, CA: Sage.

Greene, J.C. (1996). Qualitative evaluation and scientific citizenship: Reflections and refractions. *Evaluation, 2,* 277-289.

Greene, J.C., & Caracelli, V.J. (1997). Defining and describing the paradigm issue in mixed-method evaluation. In J.C. Greene & V.J. Caracelli (Eds.), *Advances in mixed-method evaluation: The challenges and benefits of integrating diverse paradigms. New Directions for Evaluation no. 74* (pp. 5-17). San Francisco: Jossey-Bass.

Greene, J.C., & McClintock, C. (1991). The evolution of evaluation methodology. *Theory Into Practice, XXX,* 13-21.

House, E.R. (1992). Response to "notes on pragmatism and scientific realism." *Educational Researcher, 21*(6), 18-22.

House, E.R. (1993). *Professional evaluation*. Thousand Oaks, CA: Sage.

Philipsen, M., & Agnew, J. (1995). Heart, mind, and soul: Head Start as a reminder of the powerful function of schools for their communities. In J.A. Hatch (Ed.), *Qualitative research in early childhood settings* (pp. 45-62). Westport, CT: Praeger.

Phillips, D.C. (1990). Postpositivist science: Myths and realities. In E.G. Guba (Ed.), *The paradigm dialog* (pp. 31-45). Thousand Oaks, CA: Sage.

Placier, M.L. (1993). The semantics of state policy making: The case of "at risk." *Educational Evaluation and Policy Analysis, 15,* 380-395.

Schwandt, T.A. (1994). Constructivist, interpretivist approaches to human inquiry. In N.K. Denzin & Y.S. Lincoln (Eds.), *Handbook of qualitative research* (pp. 118-137). Thousand Oaks, CA: Sage.

Smith, J.K. (1989). *The nature of social and educational inquiry: Empiricism versus interpretation*. Norwood, NJ: Ablex.

Stake, R.E. (1967). The countenance of educational evaluation. *Teachers College Record, 68,* 523-540.

Stake, R.E. (1975). *Evaluating the arts in education: A responsive approach*. Columbus, OH: Merrill.

Stake, R.E. (1995). *The art of case study research*. Thousand Oaks, CA: Sage.

Weiss, C.H. (Ed.), (1977). *Using social research in public policy making*. Lexington, MA: Lexington.
Whiteaker, M.S. (1996). Howard Wilson, The Teachers Academy, and School Improvement. In R.Stake et al., *School improvement: Facilitating teacher professional development in Chicago school reform. 1995-96 Final Report to the Teachers Academy for Mathematics and Science*. Champaign, IL: CIRCE.

Chapter 8

CASE STUDY METHODS

Linda Mabry

Case study is an approach to research and evaluation which emphasizes the uniqueness and situationality of a case. A case study may target a large program or phenomenon, perhaps national or international in scope, or it may be limited to one person, a single-subject study. The locus of interest may be internal, as when a researcher is motivated by a desire to learn about a specific site or issue, or external, as when an evaluator is commissioned by a client to investigate and report the merit and shortcoming of a program. The case may be examined using a variety of techniques, quantitative or qualitative—or both. In educational inquiry, case study has generally, but not always, connoted a qualitative strategy focused upon understanding the complexity, particularity, and contextuality of the case.

Every program evaluation is essentially a case study, with the program to be evaluated as the case. The program or those aspects of the program marked for investigation (i.e., the evaluand) define the bounds of the case (Smith, 1978). Concentration on the particular entity identified by these bounds makes the inquiry a case study, whether the evaluation is formative or summative; whether it attends to the interests of managers or stakeholders or policymakers; whether it highlights outcomes or processes or contexts; whether it is conducted by an internal team, an external evaluator, an accreditation panel, or a collaborative of insiders and outsiders; whether it adopts one of the so-called models such as the CIPP model (Stuf-

Advances in Educational Productivity, Volume 7, pages 155-170.
Copyright © 1998 by JAI Press Inc.
All rights of reproduction in any form reserved.
ISBN: 0-7623-0253-4

flebeam,1983), or the empowerment model (Fetterman, 1996), or participatory evaluation (Greene, 1997); whether it asserts judgments and recommendations or offers descriptive portrayal and interpretation to assist policymakers and program managers to understand or to make decisions. Almost always, a program evaluation is an *intrinsic* case study in which the program is the origin of interest, as distinguished from an *instrumental* case study in which the program is an instance of a phenomenon or issue of interest (Stake, 1994).

So, in one sense, an evaluation *is* a case study with the evaluand as its case and, in another sense, an evaluation *may also be* a qualitative case study by virtue its epistemology and methodology. This methodology may be the dominant or only strategy for the evaluation, or it may be one component of a evaluation which also includes other inquiry strategies. The purpose of this chapter is to discuss the capacity of qualitative case studies in evaluation to elevate understanding of the quality of programs and to support educational improvement.

CASE STUDY METHODOLOGY IN EDUCATIONAL INQUIRY

Epistemology, Emphases, Strategies

Traditionally, educational case study has been qualitative, interpretative, ethnographic, phenomenological, and naturalistic as described in the literature of research methodology by Denzin (1989, 1997), Erickson (1986), Lincoln and Guba (1985), Spindler (1987), Stake (1994), and Wolcott (1994). In an evaluative case study, the evaluand is studied *in situ,* with issues orienting the inquiry usually identified in negotiation with the client and with a design which is primarily *emergent* rather than preordinate. That is, the evaluation plan is purposefully flexible and adaptive in order to maximize opportunity to learn about those aspects of the evaluand which prove important, not merely those recognized at the time of the design when the evaluator is least knowledgeable about the program. This early restraint, specifying tentative inquiry issues at the outset with the expectation that these issues will be revised or replaced by more sophisticated ones as the evaluator becomes better acquainted with the program, has been termed *progressive focusing* (Partlett & Hamilton, 1976) and particularly associated with Stake's (1973) conception of *responsive* evaluation. The initial foci and their revisions or replacements which emerge during the study are informed by what is learned along the way, by the concerns and understandings of the client and program personnel and others, by the evaluator''s previous research experience, and often by the literature of the substantive field.

A case study approach to inquiry is *holistic*, based on a perspective of human interactions, such as those to be found in educational programs, as so complex and dynamic as to caution against attempts to identify and measure discrete factors or variables. The whole is understood as greater than the sum of its parts; the parts so numerous and interwoven as to elude unambiguous disentanglement and comprehensive identification; the complicated relationships among the parts confounding causal explanation; and continuous change transforming both the parts and the whole. In evaluation, this perspective works toward complex, integrated interpretations and judgments of program quality and against specification of distinct criteria and standards by which program quality can be judged.

Such inquiry is *contextual*, recognizing that social phenomena are embedded in a variety of contexts which shape the program through an intricacy of supports and hindrances. For each program, a unique array of overlapping, embedded contexts exercises influence over the nature of the program, its workings, and its outcomes. Relevant contexts are likely to include those identified by Bronfenbrenner (1979): the macrosystem or philosophical-ideological environment, the exosystem or policy context, the mesosystem or infrastructure of relationships among personnel and stakeholders, and the microsystem with its interaction-level practices and outcomes. There are many other contexts of potential importance, some primary and others peripheral, some local and others distant, some concrete and sensorily apprehensible and others subtle and recognized only through inference. There are physical, social, economic, organizational, political, legal, ethical, psychological, emotional, and intellectual contexts. In education, there are contexts of academic rigor, of pedagogy, of reform, of testing and assessment, of professional engagement and colleagueship, of teacher training and development, of facilities and resources, of tradition and reputation. There are educational contexts and community contexts. The circumstances of participants' and beneficiaries' lives, their backgrounds and opportunities, their aspirations and demands also impact expectations of the program and affect its practices.

The emphasis in this approach is on *verstehen*, on understanding rather than explanation (von Wright, 1971), the understandings of participants and stakeholders, the understandings gained and shared by the evaluator, the understandings developed by evaluation audiences. Understanding reaches deeply into epistemology. *Phenomenological*, such inquiry is oriented to perceptions of the program by a variety of persons with lived experience of it. *Constructivist*, the evaluator presumes the reality of the program is idiosyncratically constructed in the mind of each stakeholder, subjective and personal rather than objective and indisputable (see, e.g., Mabry, 1997d), and also *socially constructed*, understandings co-constructed by persons who, through their interactions and common experiences, develop intersubjective agreement regarding the nature of the program (see, e.g., Vygotsky, 1978). In an ongoing reciprocal process, the program—their understandings of it—affects their behavior, and their understandings imprint the nature of the program. This conception of case study resonates with *postmodernism*'s insistence on personal understanding as closer to truth than theory and formal explanation, the latter tending to misfocus on grand generalizations which fail to apprehend the constitutive details of actual instances (see, e.g., Rosenau, 1992;

Wakefield, 1990). It aligns with Polanyi's (1958) notion that people act on *tacit knowledge*, unarticulated understandings, more than on formal knowledge.

Deep, experiential understanding is a distinct advantage of qualitative case study which can be helpful with any educational program evaluation. Small programs are readily amenable to this approach, but case study can also be incorporated into larger program evaluations in a variety of ways. An evaluation may be a series of case studies targeting program aspects or sites, or may include such studies within a multi-method design.

Although some hold that quantitative methods dominant in the scientific paradigm are incommensurable with qualitative methods associated with the interpretive tradition (see, e.g., Lincoln & Guba, 1985), many evaluations involve both in multi-method designs intended to capitalize on the advantages of each. Not merely a matter of method selection, the interpretive mindset is not prerequisite for high quality case study. The basic strategy may be quantitative, relying on survey-based aggregations, numbers of beneficiaries served, costs, and other numerical data and indicators, with an interview component or a few case studies providing complementary experiential information about effects and practices. Published examples include the evaluation of Chicago's highly publicized state-mandated school reform (Nowakowski, Stewart & Quinn, 1992) and the Phi Delta Kappa study of at-risk students (Frymier & Gansneder, 1989). Or the emphasis may be on cases, with quantitative data included in cross-case analyses (Stake et al, 1996).

Designs

Three examples will be offered here to suggest the variety of ways qualitative case study can be used in evaluations. The first example is an evaluation undertaken for the State of Washington Office of the Superintendent for Public Instruction to determine the effectiveness of a $300,000 grant program through which the state had distributed funds to preschools and kindergartens to assist inclusion of children eligible for special services in regular classrooms. Evaluation findings were expected to inform future policy regarding inclusion. The multi-method evaluation design called for statistically analyzed surveys sent to all grantees as well as telephone interviews with the 54 project heads, site visits to 14 programs, and three case studies. The evaluation also included a review by special education and evaluation experts asked to consider the data and to critique and help refine preliminary findings (Mabry, 1995).

The case study sites were purposefully selected to illustrate a number of dimensions of interest: one preschool and two kindergartens; one exemplary program, one struggling program, and one non-inclusive program; two school settings and one setting involving a school district, a private cooperative, and Head Start facilities. While the survey, telephone interviews, and site visits provided an overview of the program which was essential and increasingly detailed, the case study data was considered by the evaluation team and review panel to have contributed critical understandings of facilitators and barriers to inclusion which were often subtly embedded in ideology, policy, and practice.

In a second example, that of the federally funded Heritage Project, case studies were conducted at implementation sites in three states. The study utilized a combination of internal and external evaluation, involving a site evaluator in each state in addition to site visits by an external evaluator (Mabry & Ettinger, 1997). Case write-ups of 12-15 pages were developed for each site in advance of cross-site analysis of the program as a whole to ensure attention to local experiences and perceptions. In this instance, careful representation of the uniqueness of each site created a tension with an overall determination of program quality. But deep understanding of each site also made possible fine-grained distinctions regarding program accomplishments and those circumstances which contributed to or obstructed success (Mabry, 1997a).

A third example, the third-year evaluation of the Chicago Teachers Academy of Mathematics and Science was designed as a series of case studies, each to focus on a different aspect of the Academy's program, a long-term plan of professional development to raise the math and science content knowledge of Chicago public elementary school teachers (Stake, Souchet, Clift, Mabry, Basi, Whiteaker, Mills & Dunbar, 1996). One case study targeted the Academy's school improvement policy, another teacher networking within associated schools, another the development of curriculum-based assessments of student achievement, and so on. The case studies ensured careful attention to targeted aspects of the Academy's agenda. Each case study raised issues for Academy personnel to consider as their program continued and evolved, moving increasingly toward a larger reform agenda.

These three examples represent different strategies for using case study in an evaluation. In the Washington OSPI study, three cases were purposefully selected to gain information about how inclusion was understood and implemented in different contexts and under different conditions. In the Heritage Project evaluation, each site at which the program had been implemented was selected for study, a comprehensive sampling. In the Chicago Teachers Academy evaluation, particular aspects of the program were identified for study, and a data-rich site selected for studying each of these aspects.

Data Collection

A case study need not be—and need not exclusively be—qualitative in approach, but many case studies employ the hallmark methods of qualitative data collection: observation, interview, and documents analysis. Educational evaluation often also involves development of survey instruments and of pre- and post-tests as well as reference to such preexisting quantitative data as enrollment and attendance figures, retention and drop-out rates, grades and standardized test scores, numbers of students eligible for special services or for free and reduced-cost meals, educational attainment of teachers, per-pupil expenditures and other budgetary information. Each source and data collection technique used offers opportunity to confirm and disconfirm data, helping to ensure the accuracy of the data set.

In addition to traditional qualitative methods, new and hybrid techniques may be developed and employed. For example, videotapes of events may be used to guide interviews, blurring the distinction between observation and interview and between data collection and analysis. Data collection, analysis, and validation similarly overlap in read-and-respond forms (Stronach, Allan, & Morris, 1996) in which data is summarized or tentatively interpreted in statements for respondents to revise so as to better represent their understandings. Technology offers such new opportunities for data collection as observations in virtual space, interviews via E-mail, and focus groups convened by means of videoconference. The power of these new methods to deepen the understanding of the quality of the case is great.

Data Analysis and Interpretation

Two approaches to the analysis of evaluation data have been prominent in the recent methodological literature. Scriven (1994) has argued for a criterial approach emphasizing professional standards within the evaluand's substantive field or, in the absence of appropriate preexisting standards, standards developed for the particular program. In this approach, data serve to document program performance along specified aspects, performance which is then compared to the standards, yielding judgments of program quality relevant to each criterion. Each criterion is weighted to ensure the primacy of those which are considered essential, protecting against the possibility of an aggregation of relatively unimportant criteria outweighing a crucial one. Analyses of program performance along specified criteria inform a *final synthesis* of overall program quality based on the standards. The process, in sum, is to identify program aspects and criteria for evaluation, to identify or develop appropriate performance standards, to judge performance of identified program aspects or criteria along these standards, and to synthesize judgments about program components into an overall determination of program quality vis-à-vis the standards.

A criterial strategy is intended to guard against invalidity arising from the distortions of subjective bias and from neglect or misprioritization of criteria in determining program quality. Its reliance on standards is respectful of substantive expertise, of consensus within a disciplined community. However, a criterial strategy is founded on two assumptions particularly troublesome in qualitative case study. First, it presumes the evaluator is capable of identifying the relevant criteria, obtaining good independent assessment of them, then recombing them in a final synthetic judgement—all with little misrepresentation or misinterpretation. From a case study perspective, such isolation and reaggregation are as likely to yield valid understandings.Aggravating this problem is the underattention to contexts usually found in a criterial approach. Decomposition and decontextualization seriously conflict with case study epistemology as described in previous pages.

Second, Scriven's (1994) a criterial approach understates the introduction of subjective bias in criteria and standards. Whether standards are selected from the field or developed for the particular program, formulated by experts or by the evaluation team, whether they are consensual or the product of a single mind, they are human constructions subject to bias. The criteria which focus data collection and analysis will emphasize articulated program aspects and exclude unarticulated aspects, ensuring attention to some criteria at the expense of others. One can hope but one cannot guarantee that the important aspects are identified. By restraining the conviction to invest in a preordinate design replete with such specification, an evaluator is better positioned to notice not only program aspects initially recognized as important but also aspects not initially appreciated. Rather than eliminating bias, a criterial approach to analysis usually assures that bias will be introduced at certain identifiable points in the evaluation process.

An holistic approach to data analysis offers a distinct alternative to the criterial approach, one more congruent with case study. It does not subscribe to decomposition of the program into separable aspects, to its disentanglement from its contexts, or to preordinate criteria. Rather, it requires critical consideration of the program as it is gradually understood in its all its interwoven layers of complexity, a process of *evolving synthesis* described by Stake and colleagues (1997) which disdains recourse to a preordinate recipe of steps. Holistic analysis is interpretive and situational, therefore difficult to describe, varying according to the perceived needs and opportunities to make meaning of the data collected. Mindful of this caution regarding the variability of actual processes, much holistic analysis could be described as a dialogic activity in which data is juxtaposed with preliminary interpretations in the constant-comparative manner described by Glaser and Strauss (1967), although an evaluation study rarely yields grounded theory which these authors conceived as the final outcome of the constant-comparative method. Data inform preliminary interpretations which, in turn, propel further data collection to confirm, elaborate, disconfirm, and refine emergent understandings.

An holistic analysis process is likely to be oriented to the issues which focused initial data collection, issues which were revised during data collection and are revised again during analysis. Data may be organized around these issues and recursively examined for relevance to themes subsumed within the issues, themes which emerge during this close examination. Micro-review of data in thematic or issue-oriented content analysis allows for repeated and increasingly minute attention to details in the data set, including those so subtle or unexpected they were little noticed during data collection. It facilitates identification of relationships among data, discovery and representation of patterns and correlations, attention to consistencies and inconsistencies and to coherence and incoherence. It yields statement and refinement of preliminary interpretations and, in their formalization, exposes them to confirmation or disconfirmation within the data set. The movement is toward formulation of findings grounded by empirical data. The effort is also an opportunity for the evaluator to select for reporting data which illustrate and provide evidentiary support for the findings. Data presentation in reports is likely to include portrayal of observed events which may help readers to develop understanding of the program through vicarious experience. It is also likely to include excerpts from interviews which express the perceptions of key participants in the program.

Interpretive inquiry requires an interpreter, the researcher or evaluator, and interpretation, a matter of informed subjective inference. The researcher or evaluator is commonly regarded as the primary instrument of data collection and interpretation. The evaluator is a connoisseur, not necessarily of subject matter as described by Eisner (1991), but a connoisseur of data and its interpretations as they contribute to understanding the case. The obvious emphasis on human judgment, although arguably no greater in qualitative than in other approaches to inquiry, opens possibilities not only for understanding but also for misunderstanding. There is potential for misrepresentation and misinterpretation, a threat to validity. Several types of efforts may protect the validity of the study (Maxwell, 1992), strategies not necessarily limited either to qualitative inquiry or to case study but clearly important to guard against the undue intrusion of subjective bias. These efforts include triangulation: explicit attempts to confirm, elaborate, and disconfirm facts and interpretations by reference to a variety of data sources including participants in different positions vis-à-vis the program, different methods of data collection (e.g., observation, interview, and documents analysis, survey, pre- and post-tests), different observers or evaluators, sustained engagement with the program over time, and analyzing data from different theoretical perspectives (see, e.g., Denzin, 1989).

Like triangulation, validation efforts serve to strengthen the accuracy of the data set and the reasonableness of interpretations and, thereby, to enhance the credibility of findings. In addition, by offering program participants a role in the inquiry, validation can promote the evaluation's utility, the likelihood that findings will be implemented. Validation can be a formal process, with participant-reviewers asked to respond to written material—selections of the data, statements of preliminary interpretations, a draft of the report. Validation may occur informally as well, with the evaluator creating conversational opportunities for participants to react to some data or to ideas about its meaning.

Lincoln and Guba (1985) described a validation process they termed *member-checking* in which a selected group of research subjects (e.g., evaluation stakeholders) representing the relevant categories of participants (e.g., students, teachers, administrators, parents, community partners) are asked to review and respond to data and emergent findings. In a more comprehensive two-staged process, not a sample of stakeholders but *each* informant might be provided opportunity to review. The first stage aims to maximize descriptive validity by offering relatively uninterpreted data for correction. Each interviewee and primary observee (e.g., the teacher of a classroom observed but not necessarily each student) is provided a write-up of the data collected from him or her, such as an interview transcript or an observation narrative, with a request that errors be corrected and additional information provided which might improve the accuracy of the data. The informant is promised that the material will not be shared further until she or he has had an opportunity to offer revisions. After a reasonable length of time has been allowed for response, a draft is produced and offered for review. This second stage aims to maximize interpretive validity by offering tentative interpretations for reaction. The draft includes selections of the data which have been validated in the first phase of review. In their perusal of the draft, informants have an opportunity to notice and comment on the evaluator's emphases and interpretations. The scope of distribution of the draft is influenced by sensitivity to ethics and politics, including such matters as possible need to protect informants' anonymity or to avoid premature distribution of interpretations in need of refinement.

In addition to review by program participants, review is expected to be undertaken internally by the evaluation team and may also be undertaken externally by review panels or colleagues. Collegial critique is formal when *meta-evaluation* (see Scriven, 1991, pp. 228-231) is undertaken to monitor and assess the quality of an evaluation. Limited resources often prevent meta-evaluation, but review by colleagues can nevertheless invigorate analysis and bolster the validity of findings by counterbalancing the subjectivity of the evaluator. Technical advisory panels may provide ongoing checks; review panels may provide critique prior or subsequent to the drafting of the report. Informally, colleagues may serve as critical friends who read, listen, comment, argue, suggest, advise. As criterial analysis suffers from bias inherent in criteria and standards and from the evaluator's imperfect ability to identify program aspects for independent assessment and then to recombine them for a synthetic determination of program quality, holistic analysis suffers from bias inherent in complex human judgment and from difficulties in ensuring rigorous critique.

For large, multi-site programs, analysis is complicated by the uniqueness of each site. Case study methodology, with its sensitivity to the singularity and situationality of phenomena, is well-suited to examination of individual sites but not to the blurring of site-specific meanings pressed to contribute to cross-site findings (Mabry, 1997a). Final judgments of program quality are not merely tallies of the quality at individual sites. In multi-site studies, analysis carries the burden of resolving or managing a palpable tension between the evaluation of the program as a whole and determinations of quality at its targeted sites.

Generalizability, as it has traditionally been construed, has long been considered limited in case study (see Campbell & Stanley, 1963; also *case-to-population generalizability* in Firestone, 1993), a limitation manifest in the particularity inherent in any evaluation of a specific program. Great caution is necessary. Generally, authors of case studies in either research or evaluation should refrain from strong, explicit claims as to the generalizability of findings. Even model programs with mandates to develop and replicate require enormous adaptation to contextual specifics; so, too, evaluations of these programs. However, the transferability of a program or its logic or processes to other settings, modified for different circumstances, may sometimes be intuitively obvious to evaluators, policymakers, and other readers. *Intuitive generalizability* can result from the deep understanding developed through the vicarious experience elicited by a case study report. Thus, evaluation by case study may support educational improvements beyond the immediate program as readers engage in *case-to-case generalizability*, recognizing applicability to another program of interest to them, or in *analytic generalizability*, recognizing applicability to a general issue or problem (Firestone, 1993). *Petite generalizations* (Erickson, 1986) circumscribed within the case promote understanding, but even the best evaluation designs are inappropriate for answering the more general question, "What works?"

Reporting

Case study reports are distinct in content and style. Where other types of evaluation reports feature data summaries, case studies typically offer substantial presentation of unaggregated, relatively incontestable or uninterpreted data, especially detailed, experiential accounts and direct quotations. Narratives, often extensive, are crafted to promote readers' understandings by encouraging vicarious experience. While the evaluator is obliged to interpret the data, generous offerings of the data help readers construct their own interpretations. The text is *readerly,* sharing meaning-making authority with readers, contrasting with the *writerly* texts of other evaluation report styles which reserve interpretive control to the evaluator (see Mabry, 1997d).

Such texts are comprehensible to a great variety of audiences, experimentally promoting readership and implementation of findings. Although a shared text may yield common notions of program quality, consensus may be obstructed by the opportunity for readers to construct idiosyncratic perspectives. Disagreement and the promotion of discussion among program participants and other stakeholders is consistent with the primary aim of a case study approach—deep understanding. But facilitation of *personal* understandings does not imply *shared* understanding. Disunity, perhaps debilitating, may be as likely as consensus.

The accessibility of the text to many audiences and the text's deliberate support for readers' development of their own conclusions have implications not only for utility but also for the credibility of the evaluator's findings. Evaluators who adopt case study methodology may offer interpretations but leave judgments as to whether programs should be continued, revised, or terminated to those whose familiarity with the program is most personal and intimate, those who must implement and bear the consequences of such decisions. Restraint of judgment, especially a withholding of recommendations, invites the question as to whether such inquiry and such reports constitute evaluation or merely portrayal. Some evaluators consider the basic task to the evaluator to be the rendering—not merely the facilitation—of judgments of program quality, sometimes including recommendations. For them, the defining task of evaluation is the formulation of judgments of quality. Lacking evaluative judgment, the inquiry is research, not evaluation, and the basic evaluation task unfulfilled. Other evaluators consider the presumption of authority by an outsider—the evaluator—to declare judgments of quality and recommendations based on brief acquaintance with the program to be inexcusably arrogant. From this perspective, the only defensible judgments of quality and plans for improvement must come from program participants whose knowledge of the program is greatest—and whose judgments can be greatly facilitated by systematic inquiry and a well-crafted report.

ETHICS AND RESPONSIBILITY

In evaluation, ethics is a matter of great seriousness, partly because results can jeopardize beneficiaries, the jobs of program personnel, the reputations of sponsors and managers, and other critical and material matters. Educational programs are embedded in micro- and macro-political scenarios characterized by competition for limited resources within and among programs. Those who have an interest in a program have an obvious interest in its evaluation. They may attempt to influence whether, how, and by whom an evaluation is conducted and to manipulate the availability of data or the dissemination of findings. Pursuant to political or personal agendae, they may work to distribute the report widely, to suppress it, or to discredit the evaluation. Political machinations are not rare in the evaluation world, where politics can intrude on professional endeavors at every point in a study's timeline. Because qualitative case study methods position the evaluator in close proximity with many stakeholders, she or he is unusually accessible to those who may desire to maneuver the evaluation. An evaluator using qualitative case study must be unusually alert to potential exploitation.

An evaluator using these highly interactive strategies must also be unusually alert to the potential for causing unnecessary damage. When, for instance, when an evaluator conducts interviews rather than administering a survey, more detailed, more personal, more revealing, and therefore more potentially injurious information may be gathered about program participants. The need for data which will provide detailed description to promote deep understanding requires balancing with the need to avoid unnecessary risk to informants. Treating vulnerable respondents as anonymous may help, but sometimes interviewees are recognizable in text because of their positions, knowledgeableness about certain program aspects, or such subtle identifiers as speech patterns. Worse, sometimes understanding a program requires understanding the perspectives and competence of key people, which forces revelation that can be embarrassing or threatening. In these situations, the integrity of the evaluation is inexorably pitted against the rights, risks, and interests of people associated with the program.

Ethical dilemmas are so common to evaluation that one might hope for guidance in the form of professional standards of practice. But formal guidance, especially *The Program Evaluation Standards* (Joint Committee, 1994) and *The Personnel Evaluation Standards* (Joint Committee, 1988), does not and cannot offer more than general principles insensitive to the particularities of immediate, practical urgencies (Stake & Mabry, 1998). Our evaluation standards (and other standards for research and professionalism) leave the evaluator with the problem of deciding which statements are relevant in a particular study, how to balance and prioritize from among the standards, and which actions might be best aligned with them. Standards do not specify appropriate response but, rather, require situational adaptation. Informal guidance may be more useful—consultation with evaluation colleagues, experience and vicarious experience through stories shared of practical and ethical problems and responses. Such stories are conspicuously absent in evaluation reports.

Nor can one preempt ethical dilemmas by refusing evaluation contracts on ethical grounds. Of course, some projects should be declined because of ethical difficulties, but most problems are not apparent at the time of engagement (Mabry, 1997b). More likely is the scenario of an evaluation terminated because of emergent or increasingly manifest problems which defy comfortable resolution. But sometimes termination may be uncomfortable, even unethical, as when an evaluation uncovers evidence of malfeasance.

An evaluator's proximity to informants can intensify the problem of positive bias (House, 1994), the propensity to produce favorable reports in order to enhance good will and the likelihood of future contracts. Personal contact can make an evaluator plainly aware of a client's distress over criticism and of powerful clients' influence. In addition to problems related to the evaluator's self-interest or self-protection, personal interaction can lead to overprotectiveness toward stakeholders. The evaluator's sympathy for hard-working or well-meaning clients, even when programs are less than exemplary, may also result in positive bias.

The issue of positive bias indexes the difficulty of balancing client expectations of an evaluation with the potentially differing interests of other stakeholders. Portrayal of multiple realities brings to light a wide variety of interests and concerns. The evaluator must adjudicate from among these interests and concerns which to highlight, which to pursue, which to champion. Some evaluators have called for emphasis on stakeholder concerns, particularly those of stakeholders who have traditionally been marginalized or whose interests are chronically neglected (see, e.g., Greene, 1997; House, 1993). Some evaluation approaches have routinized attention to certain stakeholders' concerns, making visible the evaluators' ideologies and theories. For example, Stufflebeam's (1993) CIPP model prioritizes managers' information needs in contrast to Fetterman's (1996) empowerment model which presses for capacity-building among program personnel and stakeholders. Greene (1997) promotes participatory evaluation as a means of reevaluating working relationships for the purpose of redressing imbalances of power and respect. Promotion of democratic participation is not necessarily an agenda item in case study as it is in participatory, empowerment, developmental, stakeholder-oriented, and democratic evaluation models, but case study does imply promotion of democratic process through promotion of understanding of many and by many. Information is empowering.

For an evaluation pursued via qualitative case study, discovery and articulation of the perspectives of a variety of persons involved in a program can deepen readers' understandings of the program. Evaluation audiences, in becoming more knowledgeable about a variety of experiences and perceptions, may have a fuller appreciation of the program's complexity but may also find it more difficult to reach an unambiguous determination of its quality or of improvement strategies. The challenge to the evaluator is to produce a report which facilitates understanding of this complexity without overwhelming readers or otherwise inhibiting productive change.

CONTRIBUTION OF CASE STUDY METHODOLOGY TO EVALUATION

The legitimacy of qualitative case study as an approach to program evaluation has been established in practice and in the methodological literature, where it has been championed in particular by Stake (1994) and as a member of the qualitative family by Guba and Lincoln (1985, 1989). The credibility of qualitative case study has been gained not only through articulation of its methods and rationale and through its successful use but also through recognition of the limits of quantitative alternatives in representing complex social entities (see, e.g., Madaus, Scriven, & Stufflebeam, 1987). The recombinant eclecticism of evaluation designs suggests that reliance on any single method is insupportably constraining. Case study, broadly defined as including rather than as distinguished by qualitative inquiry methods, invites the strategic adaptations good evaluation requires.

While necessitating careful efforts to guard against subjective bias and provinciality, qualitative case study methodology applied to evaluation offers unparalleled opportunity to enhance understanding through holistic portrayals strong on detail and context, illuminations which enhance vicarious experience so that understanding is not merely abstract but deeply, personally felt. Policymakers and program managers, personnel, and stakeholders are given entrée to perspectives other than their own, points of view and experiences which broaden their perceptions and deepen their comprehension of program workings and effects. They may come to understand the case, the program, because their attention is directed to the object of study rather than to a particular inquiry method, criteria, or standards.

Contributions of this type are important for the same reasons that evaluation is important. The allocation of limited resources through programs and the choice of beneficiaries should not be haphazard, nor should the value of programs be judged merely on the basis of intent or unsubstantiated opinion. Systematic collection of detailed information about a program and disciplined analysis are the tools of a rational society attempting to determine which programs contribute to the public good and how further effort might be best directed. To this general description of program evaluation, qualitative case study methodology adds distinctive depth through presentation of the multiple perspectives of insiders and through opportunities for personal, experiential understanding.

CONTRIBUTION OF CASE STUDY EVALUATIONS TO EDUCATION

Education is a complicated and difficult societal enterprise. The systematization of education by states and, increasingly, by the federal government typically neglects the fragility of the processes of learning and of cognitive development and sophistication. Consequently, educational programs often reflect a conflict between need for sensitive local, personal delivery of services and need for responsiveness to grand plans and mandates emanating from remote policymakers (see DeStefano, 1992). Except for rare instances of benign neglect or unusual understanding, this tension is rarely well resolved. Case study evaluations offer important opportunities for enhancing understanding by stakeholders, those proximal to and those distant from the program, those who set policy and control funding and those who implement or benefit. Facilitating policymakers' and clients' and stakeholders' deep understanding of a program provides a strong platform for meaningful and sophisticated program improvement.

Case studies in evaluation, undertaken as a sustained strategy, also offer a positive cumulative effect of improving education as an enterprise. Understanding one case promotes understanding of similar cases and of general issues encountered in programs. Cumulative improvements carry a promise of overall benefit to the educational enterprise derived from strengthened local programs and understandings. Improved specific understanding and improved general understanding—these promises of qualitative case studies in evaluation are among our best avenues for improving education.

ACKNOWLEDGMENT

The author gratefully acknowledges helpful comments on an earlier draft from Arthur Reynolds and Robert Stake.

REFERENCES

Bronfenbrenner, U. (1979). *The ecology of human development.* Cambridge, MA: Harvard University Press.

Campbell, D. T., & Stanley, J. C. (1963). *Experimental and quasi-experimental designs for research.* Boston: Houghton-Mifflin.

Denzin, N. K. (1997). *Interpretive ethnography: Ethnographic practices for the 21st century.* Thousand Oaks, CA: Sage.

Denzin, N. K. (1989). *The research act: A theoretical introduction to sociological methods* (3rd ed.). Englewood Cliffs, NJ: Prentice Hall.

DeStefano, L. (1992). Evaluating effectiveness: A comparison of federal expectations and local capabilities for evaluation among federally funded model demonstration programs. *Educational Evaluation and Policy Analysis, 14* (2), 157-168.

Eisner, E. W. (1991). *The enlightened eye: Qualitative inquiry and the enhancement of educational practice.* New York: Macmillan.

Erickson, F. (1986). Qualitative methods in research on teaching. In M. C. Wittrock (Ed.), *Handbook of research on teaching* (3rd ed., pp. 119-161). New York: Macmillan.

Fetterman, D. M. (1996). *Empowerment evaluation: Knowledge and tools for self-assessment and accountability.* Thousand Oaks, CA: Sage.

Firestone, W. A. (1993). Alternative arguments for generalizing from data as applied to qualitative research. *Educational Researcher, 22* (4), 16-23.

Frymier, J., & Gansneder, B. (1989). The Phi Delta Kappa study of students at risk. *Phi Delta Kappan, 71* (2, 142-146).

Glaser, B. G., & Strauss, A. I. (1967). *The discovery of grounded theory.* Chicago, IL: Aldine.

Greene, J. G. (1997). Participatory evaluation. In L. Mabry (Ed.), *Advances in program evaluation: Evaluation and the post-modern dilemma.* Greenwich, CT: JAI Press.

Guba, E. G., & Lincoln, Y. S. (1989). *Fourth generation evaluation.* Thousand Oaks, CA: Sage.

House, E. R. (1994). *Program evaluation and educational reform.* Panel discussion for conference entitled Change and Diversity: Challenges for Education, University of Illinois, Champaign, IL.

House, E. R. (1993). *Professional evaluation: Social impact and political consequences.* Newbury Park, CA: Sage.

Joint Committee on Standards for Educational Evaluation. (1988). *The personnel evaluation standards: How to assess systems for evaluating educators.* Newbury Park, CA: Corwin.

Joint Committee on Standards for Educational Evaluation. (1994). *The program evaluation standards: How to assess evaluations of educational programs* (2nd ed.). Thousand Oaks, CA: Sage.

Lincoln, Y. S., & Guba, E. G. (1985). *Naturalistic inquiry.* Newbury Park, CA: Sage.

Mabry, L. (1997a). *Critique: Contrasting approaches to data analysis in multi-site evaluations.* Paper presentation to the annual meeting of the American Evaluation Association, San Diego, CA.

Mabry, L. (1997b). *Ethical land mines in program evaluation.* Paper presentation to the annual meeting of the American Educational Research Association, Chicago, IL.

Mabry, L. (1997c). *Meta-analysis of four studies of locally developed performance assessments.* Paper presented to the annual meeting of the American Educational Research Association, Chicago.

Mabry, L. (1997d). A postmodern text on postmodernism? In L. Mabry (Ed.), *Advances in program evaluation: Evaluation and the post-modern dilemma.* Greenwich, CT: JAI Press.

Mabry, L. (1995). *Study of OSPI's three-year preschool and kindergarten inclusion grant program, methods section.* Washington State Office of the Superintendent of Public Instruction grant #34171, unpublished.

Mabry, L., & Daytner, K. G. (1997, March). *State-mandated performance assessment.* Paper presentation to the annual meeting of the American Educational Research Association, Chicago, IL.

Mabry, L., & Ettinger, L. (1997, March). *Difficulties in supporting bottom-up curriculum reform.* Paper presentation to the annual meeting of the American Educational Research Association, Chicago, IL.

Madaus, G. F., Scriven, M. S., & Stufflebeam, D. L. (Eds.). (1987). *Evaluation models: Viewpoints on educational and human services evaluation.* Boston: Kluwer-Nijhoff.

Maxwell, J. A. (1992). Understanding and validity in qualitative research. *Harvard Educational Review, 62* (3), 279-300.

Nowakowski, J., Stewart, M., & Quinn, W. (1992). *Monitoring implementation of the Chicago Public Schools' systemwide school reform goals and objectives plan.* Oak Brook, IL: North Central Regional Educational Laboratory.

Partlett, M., & Hamilton, D. (1976). Evaluation as illumination: A new approach to the study of innovative programmes. In G. Glass (Ed.), *Evaluation Studies Review Annual, 1,* 140-157.

Polanyi, M. (1958). *Personal knowledge: Towards a post-critical philosophy.* Chicago, IL: University of Chicago Press.

Rosenau, P. R. (1992). *Postmodernism and the social sciences: Insights, inroads, and intrusions.* Princeton, NJ: Princeton University Press.

Scriven, M. (1991). *Evaluation thesaurus* (4th ed.) Newbury Park, CA: Sage.

Scriven, M. (1994). The final synthesis. *Evaluation Practice, 15* (3), 367-382.

Smith, L. M. (1978). An evolving logic of participant observation, educational ethnograhy and other case studies. In L. Shulman (Ed.), *Review of research in education* (Vol. 6, pp. 316-377). Itasca, IL: Peacock.

Spindler, G. D. (Ed.). (1987). *Education and cultural process: Anthropological appraoches* (2nd ed.). Prospect Heights, IL: Waveland Press.

Stake, R. E. (1994). Case studies. In N. K. Denzin & Y. S. Lincoln (Eds.), *Handbook of qualitative research* (pp. 236-247). Thousand Oaks, CA: Sage.

Stake, R. E. (1973). Program evaluation, particularly responsive evaluation. Paper presented at conference on New Trends in Evaluation, Goteborg, Sweden. Reprinted in G. F. Madaus, M. S. Scriven, & D. L. Stufflebeam (Eds.), *Evaluation models: Viewpoints on educational and human services evaluation* (pp. 287-310). Boston: Kluwer-Nijhoff.

Stake, R., & Mabry, L. (1998). Ethics in program evaluation. *Scandinavian Journal of Social Welfare.*

Stake, R., Migotsky, C., Davis, R., Cisneros, E., DePaul, G., Dunbar Jr., C., Farmer, R., Feltovich, J., Johnson, E., Williams, B., Zurita, M., & Chaves (1997). The evolving synthesis of program value. *Evaluation Practice.*

Stake. R., Souchet, T., Clift, R., Mabry, L., Basi, M. M., Whiteaker, M. S., Mills, C. A., & Dunbar, T. (1996). *School improvement: Facilitating teacher professional development in Chicago school reform.* Urbana, IL: Center for Instructional Research and Curriculum Evaluation, University of Illinois.

Stronach, I., Allan, J., & Morris, B. (1996). Can the mothers of invention make virtue out of necessity? An optimistic deconstruction of research compromises in contrct research and evaluation. *British Educational Research Journal, 22* (4), 493-509.

Stufflebeam, D. L. (1983). The CIPP model for program evaluation. In G. F. Madaus, M. S. Scriven, & D. L., Stufflebeam (Eds.), *Evaluation models: Viewpoints on educational and human services evaluation* (pp. 117-141). Boston: Kluwer-Nijhoff

von Wright, G. H. (1971). *Explanation and understanding.* London: Routledge & Kegan Paul.

Vygotsky, L. S. (1978). *Mind in society: The development of higher mental process.* Cambridge, MA: Harvard University Press.

Wakefield, N. (1990). *Postmodernism: The twilight of the real.* London: Pluto Press.

Wolcott, H. (1994). *Transforming qualitative data: Description, analysis, and interpretation.* Thousand Oaks, CA: Sage.

PART III

ANALYSIS AND UTILIZATION

Chapter 9

STATISTICAL METHODS FOR REAL-WORLD EVALUATION

David Rindskopf

Evaluators are faced with a wide variety of statistical and design problems in their work:

- Large amounts of data;
- Complex designs, often with nesting of subjects within classes, schools, or other units;
- Missing data;
- Design of measurement instruments;
- Presenting and explaining results in simple, meaningful ways to different audiences.

Statistical methods have become both more complex and more realistic to respond to the design and data analysis problems faced by evaluators over the past few decades. New conceptual approaches and new computational methods allow researchers to model data in ways that could only be vaguely foreseen in the 1960s. These new developments have disadvantages, though: they are generally more difficult to understand, and can have untestable assumptions because evaluators may not have complete information about (for example) assignment to treat-

Advances in Educational Productivity, Volume 7, pages 173-192.
Copyright © 1998 by JAI Press Inc.
All rights of reproduction in any form reserved.
ISBN: 0-7623-0253-4

ment groups or why people drop out of studies. In this chapter I will describe new approaches to data analysis that are useful for program and policy evaluations. Because this book is meant to reach a wide audience, I will provide basic information about each method, and will give references that will allow interested readers to find more technical details. I will give a few World Wide Web addresses for specific applications; an excellent general site about statistics that seems to be relatively unknown in the United States is http://www.stats.gla.ac.uk/cti/.

The new methods span a wide range of issues relevant to evaluation. Some emphasize visual displays of data, to show important aspects of the data that might otherwise escape notice. These range from simple displays to dynamic interactive computer-generated displays. Other advances link statistical methods to philosophical underpinnings of the meaning of causation. A number of new statistical models have only been practical in the past few decades, when more powerful computing machinery became available. These include approaches to modeling selection into treatment groups (when random assignment is not performed), modeling the attrition or compliance of individuals, utilizing information about nesting (of individuals within groups, for example), and allowing for errors of measurement. In the evaluation field, many of the statistical models can be put in the broad framework of missing data models.

DATA VISUALIZATION AND PRESENTATION

Exploratory Data Analysis

For many years, a trend in statistics had been increasing reliance on sophisticated mathematical procedures and on inferential statistics, and less reliance on descriptive statistics and graphical methods. John Tukey practically single-handedly reversed this trend with his concept of exploratory data analysis (Tukey, 1977). Now statistics texts are more likely to present coverage of exploration, description, and inference, rather than just inference. Naturally, this is as important for evaluation as for any other applied area. For a tutorial on the application of exploratory data analysis in evaluation, see McGaw (1981); other overviews are in Leinhardt and Wasserman (1978), and Leinhardt and Leinhardt (1988). An excellent example of the application of these techniques in a real evaluation is Abt and Magidson (1980).

Modern Graphical Tools

The development of better hardware and software for graphics, especially on personal computers, has led to an increase in the availability of graphical tools for data analysts. As one example, if one has several continuous variables, a matrix of scatterplots of each variable versus every other variable can be produced by most statistical packages. This allows the researcher to examine a large number of relationships by looking at one page (or screen), and perhaps notice patterns that might otherwise be difficult to detect.

Other methods combine intense computation with graphics, such as density and regression smoothers. For example, suppose the relationship between a continuous outcome Y and a continuous predictor X may not be linear, but also may not be described by a simple relationship (such as a polynomial). A plot of Y versus X might not clearly indicate the nature of the relationship, which is a conditional mean of Y at each value of X. A smoother would calculate the mean (or median or other central tendency measure) of Y for a "window" around each value of X, to make sure that there are enough values to compute the mean with stability. The plot of these means versus X forms a continuous line, which can be examined to determine the shape of the relationship. A number of books illustrate modern graphical methods; a "classic" one is Chambers, Cleveland, Kleiner, and Tukey (1983). For regression graphics, see Cook and Weisberg (1994), which includes a disk with a computer program to do all of the graphics discussed in the book.

Dynamic Graphics

With greater computer power, one can invent new graphical techniques that allows the researcher to "travel" through a data set in multidimensional space, or rotate a data cloud in any direction. One can plot a set of data, and by using a mouse to move a marker, watch how the nature of the relationship changes as the transformation of either the X or Y axis is changed. Two sources for these techniques are Cleveland and McGill (1988) and Cook and Weisberg (1994).

Presentation of Results

While most work in graphics is aimed at helping the researcher understand data, some is aimed at helping the researcher explain the results to the final audience: administrators, legislators, judges, the press, and the public, to name a few. The construction of tables and graphics to maximize clarity and informativeness is discussed in Ehrenberg (1975, 1977), Tufte (1983), Tufte (1990), and Wainer (1997).

LATENT VARIABLE MEASUREMENT MODELS

Most modern psychometric theory revolves around the idea of latent variables. These are the hypothesized underlying constructs, of which observed measures are only imperfect indicators. The latent variables may be categorical (e.g., stage of development) or continuous (e.g., level of mathematical knowledge). Further, the observed variables may be either categorical or continuous (or some combination). Different techniques have been developed to handle different combinations of types of observed and latent variables. Some of the common methods are described below. A more complex model of causal effects that includes factor analysis in the measurement part of the model is described in the "Missing Data" section. One can, in fact, conceptualize all of the methods in this section as missing data models, because the values of any latent variable are, by definition, missing for all individuals.

Factor Analysis

Suppose that you have measured 200 high school students on six tests: spelling, grammar, literature, arithmetic, algebra, and geometry. On the basis of your knowledge of these areas, you might hypothesize that there are two broad academic abilities that are being measured: verbal ability, and mathematical ability. Or you might believe that there is one broad, general, academic ability that lies behind all six measures. Either of these hypotheses can be made very precise, and are examples of factor analysis models. In general, these are models for continuous latent variables (factors) and continuous observed variables, but recently much progress has been made in extending the models to include dichotomous and ordered categorical variables as well. The usefulness of factor analysis is that it can represent a great simplification if one only cares about the broad general abilities, because then one need only keep track of one or two factors (in this example) rather than six observed variables. A good introduction to modern methods of confirmatory factor analysis is contained in Long (1983). Extensions of factor analysis, discussed below, are also useful because they allow the estimation of treatment effects, controlling for unobserved covariates rather than merely controlling for observed variables.

Latent Class Models

Latent class models are similar to factor analysis in many ways, except that both the latent variables and the observed variables are categorical. For example, if one gives four items designed to assess whether a child can conserve volume, a Piagetian might hypothesize that there are two types of children (two latent classes): those who can conserve, and those who cannot. But errors of understanding or expression, or lucky guesses, might add error to the observed responses. Latent class analysis can help decide whether the observed patterns of right and wrong responses are consistent with the simple model that there are only two kinds of children with respect to conservation of volume. McCutcheon (1987) is a good source on the basics of latent class analysis.

Item Response Theory (IRT)

Psychological scales are often based on summing scores of a number of individual items. Often little or no thought is given as to whether summing scores is a reasonable operation. Psychometric theory, however, has advanced considerably in the last three decades, providing new tools that can be valuable to evaluation researchers. All involve latent variable models, and most are related to factor analysis models; some are also related to loglinear or latent class models. Most are based on the notion that each dimension that is being measured is an underlying continuum, and that observed responses are a probabilistic function of a person's place on that continuum. While the theory was originally developed with test scores in mind, it also can be fruitfully applied to psychological constructs such as locus of control, self-efficacy, introversion, and so on.

The simplest and best-known item-response theory (IRT) model was developed by Georg Rasch. It has primarily been applied to dichotomous items (scored 0 or 1), but models and corresponding computer programs exist for ordered categorical responses (e.g., Likert items), and for more complicated underlying statistical models.

There are two primary benefits of using IRT models in evaluation. One is to develop a scale with more reasonable measurement properties. A second is to facilitate vertical equating, so that pre- and post-test scores can be on the "same scale," even though they contain items of different difficulties. For example, one cannot always use the same reading test for third and fourth graders, but using IRT to equate tests allows both third and fourth graders to be measured on the same scale. Hambleton, Swaminathan, and Rogers (1991) provide an excellent introduction to IRT.

CONCEPTUAL FRAMEWORKS FOR CAUSAL MODELS

Evaluators constantly face the problem of assessing whether a program has had an effect, and if so, estimating the size of the effect. Nonrandom assignment to conditions is the rule rather than the exception, and even randomized studies face problems due to attrition and noncompliance. Evaluators are therefore very concerned with methodological issues surrounding causal inference. Several statisticians have contributed to this area recently; the work of those whose writings are most useful for evaluation is summarized in this section. A good overview of statistical models related to causal inference in the social sciences can be found in Sobel (1995).

Rubin's Causal Model

Rubin (1974, 1977, 1978) has developed a useful framework for conceptualizing studies involving group comparisons. This framework helps make explicit (i) assumptions involved in analyses, (ii) different definitions of causal effects that might be important, and (iii) which questions about causation are valid, and which are not. In this conceptualization, everyone has two (or more) potential outcomes, only one of which is observed: the outcome that would be observed if they received the experimental treatment, and the outcome that would be observed if they received the control treatment. Causal questions concern the differences between the responses that we would observe under these two conditions.

When viewed in this way, the major statistical problems are easily seen to be missing data problems: We are missing data on performance under control conditions for those who receive the experimental treatment, and vice versa. Extensions of the basic model deal with problems such as noncompliance with researchers' instructions, attrition, and studies in which choice of treatments is at least partly (or totally) determined by the subjects. The basic theory and implications are described in a chapter by Holland and Rubin (1983).

Rosenbaum's Work on Observational Studies

Rosenbaum (1995) gives a thorough conceptual and statistical basis for the analysis of data from observational studies. One of the most useful parts of his book is the extensive discussion of the use of sensitivity analysis to give plausible bounds on the amount of bias due to omitted variables in a model. Because analyses of observational studies usually involve assumptions that cannot all be empirically verified, sensitivity analysis is an important tool to determine how much the results would have changed if the assumptions are wrong. Rosenbaum demonstrates how to quantify the amount of hidden bias due to omitted variables in an analysis, and how to measure their effect on the outcome. He shows studies in which no plausible amount of hidden bias would change the results (smoking and cancer, for example), and other studies in which minor amounts of hidden bias would change the conclusions of the study. In the smoking and lung cancer case, for example, one would ask how strong would the effect of an omitted variable have to be in order to reduce the relationship between smoking and lung cancer to zero. Rosenbaum shows that an unobserved covariate would have to increase the odds of smoking in lung cancer patients by at least six-fold over the odds of smoking in controls in order to reduce the relationship between smoking and lung cancer to nonsignificance. It is unlikely that researchers have overlooked any variable with such a strong effect, so we conclude that the relationship is quite robust.

Another important area for evaluation in Rosenbaum's work is his discussion of coherence, which refers to the overall pattern of results in a study. The outcome of a study shows coherence if most of the results are in accordance with what theory would predict. If there are several outcome variables that should be affected by treatment, then all (or most) should be shown to be related to treatment. If there are other variables measured at the same time as outcome variables that should not be affected, then groups should not differ on these variables. If a large number of sites are involved, then most or all should show the effects. A number of these and related issues are discussed in the context of evaluation by Rindskopf and Saxe (1998). The case they discuss concerns the evaluation of a program to reduce the use of drugs, alcohol, and tobacco. Some dependent variables (e.g., binge drinking) might reasonably be expected to be more easily affected than others (e.g., cocaine usage). If cocaine usage dropped precipitously but binge drinking did not, one might wonder about the reliability of the data. Similarly, because the program was implemented in 14 sites, if the effect only occurred in one or two sites (especially if these were not the sites that had the best implementation), one would question the data. In each of these cases, one can informally display the overall pattern of results, as well as conduct formal statistical analyses to see if any patterns exceed chance expectations.

MISSING DATA

Evaluators encounter missing data much more frequently than do laboratory researchers, and the processes causing missing data are often much more likely to cause bias in evaluations than in laboratory studies. Modern statistical approaches to missing data are based on explicit probability models for the missing data process, and are usually based on methods involving either maximum likelihood estimation or Bayesian methods (both of which involve specification of the likelihood function). Older methods (such as case-wise or list-wise deletion, or mean substitution) are primarily ad hoc procedures, and can give very misleading results.

Missing data is a very broad area; in fact, much broader than might be supposed. It encompasses many of the situations discussed in this chapter. For example, consider a study with two possible treatment conditions (say, experimental and control). In Rubin's conceptual model, each person has a potential score that he or she would receive in each of the treatment conditions. But only one of these is observed, because the person either receives the experimental or the control treatment, but not both. Therefore, the score they would receive under the other condition is missing. All latent variable models (such as factor analysis, latent class analysis, cluster analysis, and item response theory) can be viewed as situations involving missing data: everyone is missing the information about their status on the latent variables (i.e., factor scores, latent class membership, true cluster to which they belong, etc.). Noncompliance is a missing data problem, because we do not observe what outcome would have occurred if the person had complied with instructions. In sum, missing data models provide a unifying framework for thinking about many causal inference problems in evaluation.

In the next section, I will discuss general approaches to missing data. Following that are sections devoted to specialized issues in missing data.

General Approaches to Missing Data

Modern theory classifies missing data problems into three broad categories. Data are *missing completely at random* (MCAR) if being missing is unrelated to the observed variables. This would occur, for example, if data were missing because a measuring instrument failed at random times. Data on a variable Y are *missing at random* (MAR) if missingness on Y is related to another variable X, but not to Y. For example, if X is used to select people into a program, and Y, the outcome, is only observed for people whose X score exceeds some value, then Y values are MAR. Data are *not missing at random* if missingness on Y is related to the value of Y that would have been observed. For example, if people with high incomes are more likely to refuse to respond to a question about income in a survey, the data are not MAR; in this case, nonresponse is said to be nonignorable.

Two excellent overviews of modern methods for dealing with missing data are the chapters by Little and Rubin (1990) and Little and Schenker (1995). The best detailed work on missing data is the book by Little and Rubin (1987). References appropriate for particular problems are given in other sections of this chapter.

Few general statistical packages implement modern missing data methods, though some are making strides in the right direction, and others are about to become available. BMDP has for several years had routines that help investigate patterns of missing data, and use maximum likelihood techniques to estimate means, variances, and correlations among continuous variables (though not standard errors). SPSS has recently implemented some similar routines, and also has procedures to impute missing values. Neither package has a straightforward procedure to estimate standard errors for the statistical analyses that would result from using incomplete data. No statistical package now in common use has good routines for handling missing categorical data, though it appears that this will change shortly.

Some statistical models are so complex that they cannot be easily programmed for solution in a general way. This is especially true of data sets with general patterns of missing data. New numerical methods have been developed to deal with these complex situations. One promising method deserves special attention because a computer program, called BUGS, has been written to enable researchers to apply these methods. Theory and applications of this method, called Markov Chain Monte Carlo, are in Gilks, Richardson, and Spiegelhalter (1995); the program manual is Spiegelhalter, Thomas, Best, and Gilks (1995); and a web site from which the program and manual can be downloaded is http://www.mrc-bsu.cam.ac.uk/bugs.

Models of Group Membership

A number of approaches to causal inference in nonrandomized studies involve modeling the process by which people are selected (or self-selected) into different treatment groups. If one can model this process, then one can obtain unbiased causal effect estimates. The most well-known example is the regression discontinuity design (Thistlethwaite & Campbell, 1960) popularized by Donald T. Campbell. In this design, a continuous variable (such as the score on a qualifying test) is used to decide which people will get a special treatment (e.g., being named a National Merit scholar). The outcome might be grade point average during the first year in college. If the "treatment" had an effect, then the regression of grade point average on the qualifying score should show a jump, or discontinuity, at the cutoff point on qualifying scores.

The regression discontinuity design, while potentially useful, is seldom implemented, and so in real-life evaluations we are seldom able to find a variable (or even a set of variables) that perfectly accounts for group assignment. New statistical methods have been developed that will give correct causal effect estimates under some circumstances. Three major strands of methods have been developed for this type of model; their similarities far outweigh their differences: Selection modeling, relative assignment variable method, and propensity scores.

Selection modeling, popularized by Heckman (e.g., see Heckman & Robb, 1986), uses a parametric model for the probability of being in the experimental (rather than the control) group. The simplest formulation of the model has two equations. The first equation models the probability of being in the experimental group, and is generally either a probit or logistic regression. The second equation models the dependent variable as a function of whatever variables (including actual group membership) are thought to influence the outcome. Unlike some models, one can allow for correlation between the error in the two equations.

The *relative assignment variable* approach also models the probability of being in the experimental group, but generally in a nonparametric fashion (e.g., Trochim, 1984). Suppose that a variable (e.g., family income) is thought to be important in determining whether children get compensatory education. For each child's income, we would find all of the children in the study whose family income was "near" (e.g., within $3000) that amount. Then we would find the proportion of those children who were assigned to compensatory education. An advantage of this method is that it does not make any assumptions about the form of the relationship between family income and probability of receiving compensatory education. Disadvantages are that it becomes more difficult to apply with more than one predictor of group membership, and it cannot handle correlated errors between the assignment and outcome equations.

Propensity scores have similarities to both of the above methods. The propensity refers to the probability of being in the experimental group. It can be modeled in any way the analyst believes to be reasonable, whether parametric (e.g., logistic regression) or nonparametric. The propensity score is then included in the outcome equation along with the actual treatment assignment variable.

A major (but unavoidable) problem with all of these methods is the need to make assumptions that are usually untestable. If these assumptions are false, then the estimates of treatment effects will be biased. More generally, this is a problem for most statistical analyses involving missing data. By making the assumptions explicit, and using information from subject-matter knowledge in a particular context, one can sometimes determine whether these assumptions appear reasonable. Sensitivity analysis is another possibility: One can investigate how much the answer would change if the assumptions were wrong by various amounts, or if different sets of assumptions were made.

A nontechnical account of selection modeling, relative treatment assignment, and propensity scores is Rindskopf (1986). Selection modeling is described in detail in Maddala (1983). A classic paper on selection modeling is Barnow, Cain, and Goldberger (1981). The mixture modeling approach is described and compared with selection modeling in Glynn, Laird, and Rubin (1986), Little and Rubin (1987), and Heckman and Robb (1986). LIMDEP (Greene, 1995; see also http:// wuecon.wustl.edu/limdep/limdep.html) implements many procedures that are useful with sample selection, as well as with similar problems such as truncation, censoring, and attrition.

Latent Variable Causal Models

As mentioned above, many constructs in psychology and education are not directly observable. Any observed measure of these latent constructs has measurement error, and by definition data are missing on the desired underlying construct. Confirmatory factor analysis models and structural equation models (Jöreskog & Sörbom, 1979) have expanded the scope of measurement error models well beyond traditional models in test theory. While these models have many applications in evaluation, the one with the most potential for removing bias in estimating treatment effects was proposed by Sörbom (1978; reprinted in Jöreskog & Sörbom, 1979). For many years, Donald T. Campbell had pointed out that regression artifacts could bias nonequivalent control group studies when the control variables were measured with error (see, e.g., Campbell & Erlebacher, 1970). Sörbom's methodology was the first to provide a general solution to this problem. He showed that with multiple measurements of the control construct, one could correct for measurement error in the control variables. Bentler and Woodward (1979) and Rindskopf (1981) give overviews of the use of structural equation models in evaluations.

How does Sörbom's method compare with the current work on selection modeling, propensity scores, and other related methods? Sörbom's method (and analysis of covariance in general) corrects for group differences in variables that are thought to affect the outcome. The assumption is that there are two populations (control and experimental), and the control variables tap all of the sources of pre-existing differences in factors that might affect the outcome. Selection modeling and propensity scores, on the other hand, are controlling for factors that relate to which group a subject is in; these factors may or may not be those that are strongly related to the outcome. Each model has its assumptions, and in most cases, one will never know which approach is correct; in any application the analyst must determine which set of assumptions seem most reasonable. Unfortunately, they do not always give the same results, as was shown many years ago by Cronbach, Rogosa, Floden, and Price (1977) in a paper that was widely circulated but never published.

Survival Analysis

Suppose that a program attempts to reduce recidivism among juvenile offenders. If one follows treated and control subjects for a year, the outcome could be whether or not they were rearrested. The analysis would then be straightforward, involving logistic regression. Suppose instead that one felt information would be lost this way, because data are available about when subjects were rearrested. Why not consider length of time to rearrest, which should be more sensitive as an outcome measure? That would lead to an analysis for counts (Poisson regression for number of months to rearrest), or possibly the outcome would be treated as a normal variable, and multiple regression would be used. The problem (from an analytic standpoint) is that not all subjects will have been rearrested (we hope); their time to rearrest is said to be censored. Methods originally developed for biostatistics (length of time until death or reoccurence of disease) and reliability analysis (length of time until failure of a device) can be usefully applied in these cases. Useful references on the application of these methods in the behavioral sciences include Singer and Willett (1991, 1993).

Noncompliance and Attrition

The standard method for dealing with noncompliance is to ignore it. This results in an "intent-to-treat" analysis, in which everyone in the experimental group is compared to everyone in the control group, regardless of whether those in the experimental group actually got the treatment (or whether those in the control group actually did not get the treatment). This provides a conservative estimate of the treatment effect, but may underestimate the effect of treatment had everyone complied with the researcher's design. (Note that using only those in the experimental group who actually complied, and comparing them with the whole control group, is never correct, because compliers will usually differ from noncompliers in their outcome even if untreated.)

Often, however, information is available on compliance in either the experimental group, or in both the experimental and control groups. In these cases, new methods are available that will allow the estimation of treatment effects among compliers. Some assumptions are needed for these procedures to work correctly; in most cases, random assignment to groups is necessary. The first to discuss these methods in an evaluation context was Bloom (1984). His article clearly spells out a simple method which, while not as statistically efficient as maximum likelihood, is easy to understand and implement. Another instructive article is Sommer and Zeger (1991), who show how to compute effect estimates when the dependent variable is categorical. Little and Yau (in press) make explicit the assumptions necessary for these methods to work, and show how to use maximum likelihood to estimate the parameters.

GENERAL MODELS FOR RECTANGULAR (FLAT) DATA FILES

Much of classical statistics has developed around data sets that can be represented as rectangular (flat) data files. These files are ordinarily structured so that the rows represent individuals, and the columns represent variables. This approach is embodied in all of the traditional computer packages for statistical analysis, and is explicit in packages such as SPSS, which structure the data in the form of a spreadsheet. Developments in the analysis of flat files have generally proceeded in the direction of expanding the types of variables that can be analyzed, and fitting all of the analyses into one common framework.

General Linear Model Approach

By now most statisticians use the general linear model (GLM) approach to a wide variety of data analyses. Before the GLM approach, a *t*-test looked different from an ANOVA, which looked different from regression, which looked different than ANCOVA. Now, they are all seen as special cases of multiple regression. The "classic" approach to general linear models concerns using multiple regression, with special coding to handle categorical predictors. Current texts adopting this approach include Lunneborg (1994) and Judd and McClelland (1989), among many others. One natural extension is multivariate analysis of variance (MANOVA), where there is more than one (continuous) dependent variable.

Categorical Data Models

Many outcome variables in evaluation are categorical, most notably success or failure to attain some goal, but also graded responses (poor, good, excellent outcome). Modern methods for analysis of categorical data were developed by Goodman (1978) and Bishop, Fienberg, and Holland (1975). These include loglinear models for describing relationships among a number of categorical variables, logit models for categorical dependent variables, and latent class models for unobserved categorical variables. Models for observed variables are discussed in detail by Agresti (1990); logistic regression by Hosmer and Lemeshow (1989), and latent class analysis by McCutcheon (1987). Many of these methods can be placed within the context of generalized linear models (see below).

Generalized Linear Models

Eventually a large number of models were seen to be special cases of what is now termed the Generalized Linear Model (GLM; yes, the same acronym as is sometimes used for general linear models). Generalized linear models are extensions in two major ways. First, the distribution of errors is not limited to the normal, but includes a large number of distributions (such as the Poisson and binomial) that allow modeling of a wider variety of dependent variables. For example, Poisson regression and logistic regression are special cases of GLMs. Second, it is more convenient to represent the dependent variable in a transformed state for many models, thereby reducing a nonlinear model (for the original dependent variable) to a linear model (for the transformed variable). For logistic regression, the link (transformation) is the logit (log odds); for loglinear models, it is the logarithm.

The literature on these models is vast, and includes introductory (e.g., Dobson, 1983) and advanced works (e.g., McCullagh & Nelder, 1989), as well as books demonstrating the use of computer programs for GLM (e.g., Aitkin, Anderson, Francis, & Hinde, 1989).

MODELS FOR COMPLEX DATA STRUCTURES

A wide variety of data sets from evaluations involve data with more complicated structures than assumed by traditional methods. Many can be viewed as special cases of multilevel models. Two excellent summaries of multilevel model approaches are Bryk and Raudenbush (1992) and Goldstein (1995). Corresponding to these books are the computer programs HLM and ML-n developed by their authors (in collaboration with others). The ML-n web site (http://www.ioe.ac.uk/multilevel/mln.html) contains a tutorial, and has pointers to web sites for other multilevel programs. Major statistical packages such as SAS and SPSS have recently added multilevel modeling routines, and there are a number of other specialized programs (VARCL, MIXREG, and MIXOR are examples) that have advantages in certain situations. Articles on the use of multilevel models in evaluation include Seltzer (1994), and Osgood and Smith (1995). Some typical situations in which these models are applied are given below.

Nested Data Structures

More evaluators are coming to the realization that their designs involve multiple levels of random variables. In community-based evaluations (Murray & Wolfinger, 1994; Rindskopf & Saxe, 1998) people are nested within communities. In many educational evaluations, students are nested within classes, schools, districts, or states. It has long been known that ignoring this nesting would generally lead to rejecting the null hypothesis too frequently, but only recently have useful statistical models existed for such data, and computer programs have become widely available for data analysis.

Longitudinal (Repeated Measures) Data

Another application of multilevel statistical models in evaluation is the analysis of repeated measures (longitudinal) data. Traditional methods for repeated measures analysis have problems with missing data; while laboratory studies generally do not have much missing data, field studies frequently cannot locate respondents at one or more times scheduled for measurement. Furthermore, while the research design may call for measurement of everyone at the same time or ages, this may not occur due to logistical problems in the management of a study. Current thinking about longitudinal data concentrates on fitting growth curves for each person, and determining how much the characteristics of those curves vary across people. For example, one could start by seeing whether a straight line fits the data for each person. Each person's growth would then be summarized by a slope and an intercept (with the point of origin carefully considered). These slopes and intercepts might then be modeled as a function of characteristics of the individuals, such as gender, household income, and so on. Multilevel models handle these cases easily, and are described in Goldstein (1995) and Bryk and Raudenbush (1992). They are extremely useful for repeated measurements that are too short for the usual time-series methods of analysis. Models are also available for nonlinear growth functions, which are very common in some fields, but have not been applied in evaluation.

Meta-Analysis

When an evaluation is conducted to evaluate the same program in a number of sites, or a large number of evaluations are conducted to evaluate similar programs, one is frequently interested in combining the information from the various sites or evaluations. One might want to know what the average effect size is, and whether the size of the effect shows variation among the sites (or evaluations) beyond what would be expected due to sampling variation. If substantial variation exists, one might want to know what characteristics of the sites (or programs) are related to variation in the effect size. The general area of meta-analysis deals with these issues. Because one can view a meta-analysis as the analysis of data collected on individuals nested within sites or studies, the methods of multilevel models can be applied. Thus, the statistical conduct of a meta-analysis is just a special case of multilevel models (although meta-analysis also has measurement and design issues), and is described in detail in Bryk and Raudenbush (1992). Rindskopf and

Saxe (1998) discuss some issues in the application of these models to the evaluation of community-based programs.

Improved Estimates Using Empirical Bayes

Evaluators must sometimes gives estimates of some quantity of interest for each of a number of institutional units. For example, in education one might want to examine the percentage of students passing an examination in each school in a school system. In medicine, one might want to determine the mortality for a given operation in each hospital in a system, as New York State has done for coronary artery bypass graft (CABG) operations. Using techniques now generally referred to as empirical Bayes (EB), one can estimate these quantities more accurately than by using the separate data from each unit. In EB, one "gathers strength" by using data from other units (e.g., other schools or hospitals) to estimate the quantity of interest, by assuming that all of the units have something in common. A good introduction to the logic of this process is Efron and Morris (1977).

SPECIAL CONCERNS IN STATISTICAL ANALYSIS

A few topics with direct relevance for evaluation fit in special categories outside of the major areas of statistics. Some of the following have also been mentioned above, but deserve special attention.

Sensitivity to Assumptions

For many of the classic statistical models, it was a relatively easy matter to test critical assumptions such as normality and homogeneity of variance. Newer models make more assumptions that are untestable. If assumptions are untestable, how can we have confidence in the results of an analysis? As mentioned previously in the discussion of Rosenbaum's work, one approach is to determine how wrong the assumption would have to be in order for the result to change. This approach is called sensitivity analysis, because it investigates how sensitive the results are to the assumptions.

An early example in the evaluation literature is Rindskopf (1978), who tested the sensitivity of reliability-corrected ANCOVA to various possible estimates of reliability. A good source on general principles of sensitivity analysis in observational studies is Rosenbaum (1995).

Multiple Statistical Tests

Researchers generally encounter the term "multiple comparisons" when studying one-way analysis of variance. In that context, the problem of conducting multiple statistical tests arises in comparing the means of each pair of groups in a study. The problem is a more general one, however, and often arises in the context of evaluation. Many evaluations involve collecting data on a wide variety of outcome measures, several groups that received different treatments, and sometimes subgroups (based on sex, race, age, pretest scores, and so on). If every group and subgroup comparison were done for each outcome measure, a large number of statistical tests would be done, leading to the problem of several tests being likely to be significant due to chance.

Tukey (1977), who brought the multiple comparison problem to the attention of researchers in its original context many years ago, was also one of the first to point out the problem in the context of evaluation (in this case, the evaluation of medical treatments). The usual solution to these problems has been to adjust for the number of comparisons that are done, using any of a number of techniques. But in the evaluation context, more can be done. For example, one can simply note whether the significant effects are all in the same direction (which would be unusual if only chance were operating) or divided between both positive and negative effects. One can use multivariate methods (factor analysis, multivariate analysis of variance) to reduce the dimensionality of the outcome variables, and thus reduce the number of statistical tests. See Rindskopf and Saxe (1998) for more details on using these and other methods in the evaluation context.

De-Emphasizing Significance Tests

A recent trend in some applied statistical areas is to de-emphasize the role of statistical hypothesis tests, and to supplement (or replace) hypothesis testing with interval estimation. Some journals have gone so far as to ban significance testing. Certainly confidence intervals provide all of the information needed for hypothesis testing, and more. They give a range of plausible values for a parameter, and in doing so provide information about the precision with which the parameter has been estimated. The advantages of confidence intervals are even greater when outcome variables are on scales that have some meaning to the reader.

A related issue concerns effect size: How big is the observed effect? Classical applied statistics has always emphasized that an effect must not only be statistically significant, but be large enough to be practically important, in order to take it seriously. The American Psychological Association (1994, p. 18) tells its authors to provide effect size indicators, and has established a task force on statistical inference to provide recommendations about the use of hypothesis tests, effect size indicators, confidence intervals, and related issues. Several books and special issues of journals have addressed these issues; an example is Harlow, Mulaik, and Steiger (1997).

Simple Presentation of Complex Results

An issue of great importance in the real-world evaluation is how to present results to an audience that is not interested in research design or statistical analysis. The evaluation audience doesn't care how hard it was to design the study, collect the data, or do the analysis (except for the impact on their budget and time schedule); all they care about are the results. They don't understand logits, variance components, factor loadings, or Markov Chain Monte Carlo methods, and they don't want to.

The literature on presentation of results is scant (see Ehrenberg, 1975, 1977, for excellent advice on structuring tables, and Tufte, 1983, 1990, and Wainer, 1997 on graphics). However, a few simple rules may help. First, translate all results into a scale that the audience understands. For example, always translate results of a logistic regression into proportions, even though technically this is problematic. An effect on a logit scale might correspond to a 10 percent difference at one point on the scale, but a 7 percent difference at a different point on the scale. Nonetheless, one must accept some slight technical slippage in order to accomplish any communication with a nontechnical audience.

Second, if a large number of statistical tests is done, develop summary tables of the results, so that any patterns become apparent. For example, an evaluation of a drug and alcohol program involved 12 treatment sites and 7 major outcome variables. A 12 x 7 table with a plus (+) for positive, minus (−) for negative, and 0 for no significance provides the audience a good summary of a large number of statistical tests. It will also show whether effects are consistent (across sites or across outcomes), and thus more believable; or inconsistent, and thus more likely due to chance.

Third, translate all important results into verbal statements, even if they are only approximately correct. Excellent examples are in Bryk and Raudenbush (1992), as well as other articles of theirs. They explain each parameter of complex multilevel models in the language of the content area of the study they are analyzing.

CONCLUSIONS

Statistical methods have developed considerably in the past two decades. Current models are more realistic, more complex, and require more intensive computation than previous models. At the same time, many researchers have realized that results must be interpreted and presented so that relevant policymakers can properly apply them. While complex models provide greater realism, they can also involve a number of additional assumptions, some of which may not be easily testable. The need for such assumptions is especially critical in situations involving missing data; as indicated, this is a much broader area than considered by most researchers, and includes latent variable models, nonequivalent group designs, attrition, and noncompliance as well as the traditional "missing observations." The new conceptual approaches and statistical methods that are now available provide improved tools for evaluators with complex designs and data sets.

ACKNOWLEDGMENTS

Thanks to Dr. Arthur J. Reynolds and Dr. Laurie Hopp Rindskopf for comments on an earlier version of this chapter.

REFERENCES

Abt, W. P., & Magidson, J. (1980). *Reforming schools: Problems in program implementation and evaluation.* Beverly Hills, CA: Sage.

Agresti, A. (1990). *Categorical data analysis.* New York: Wiley.

Aitkin, M., Anderson, D. Francis, B., & Hinde, J. (1989). *Statistical modeling in GLIM.* Oxford: Oxford University Press.

Americal Psychological Association. (1994). *Publication manual of the American Psychological Association* (4th ed.). Washington, DC: Author.

Barnow, B. S., Cain, G. G., & Goldberger, A. S. (1981). Issues in the analysis of selectivity bias. Pp. 43-59 in W. E. Stromsdorfer & G. Farkas (Eds.), *Evaluation studies review annual* (Vol. 5). Beverly Hills, CA: Sage.

Bentler, P. M., & Woodward, J. A. (1979). Nonexperimental evaluation research: Contributions of causal modeling. Pp. 71-102 in L. Datta & R. Perloff (Eds.), *Improving evaluations.* Beverly Hills, CA: Sage.

Bishop, Y. M. M., Fienberg, S. E., & Holland, P. W. (1975). *Discrete multivariate analysis: Theory and practice.* Cambridge, MA: MIT.

Bloom, H. S. (1984). Accounting for no-shows in experimental evaluation designs. *Evaluation review, 8,* 225-246.

Bryk, A. S., & Raudenbush, S. W. (1992). *Hierarchical linear models: Applications and data analysis methods.* Newbury Park, CA: Sage.

Campbell, D. T., & Erlebacher, A. (1970). How regression artifacts in quasi-experimental evaluations can mistakenly make compensatory education look harmful. In J. Hellmuth (Ed.), *Compensatory education: A national debate. Vol. 3: Disadvantaged child.* New York: Brunner/Mazel.

Chambers, J. M., Cleveland, W. S., Kleiner, B., & Tukey, P. A. (1983). *Graphical methods for data analysis.* Belmont, CA: Wadsworth.

Cleveland, W., & McGill, M. (1988). *Dynamic graphics for statistics.* New York: Chapman & Hall.

Cook, R. D., & Weisberg, S. (1994). *An introduction to regression graphics.* New York: Wiley.

Cronbach, L. J., Rogosa, D. R., Floden, R. E., & Price, G. G. (1977). *Analysis of covariance in nonrandomized experiments: Parameters affecting bias.* Occasional paper, Stanford University, Stanford Evaluation Consortium.

Dobson, A. J. (1983). *An introduction to statistical modelling.* London: Chapman and Hall.

Efron, B., & Morris, C. (1977, May) Stein's paradox in statistics. *Scientific American,* 119-127.

Ehrenberg, A. S. C. (1975). *Data reduction: Analyzing and interpreting statistical data.* New York: John Wiley.

Ehrenberg, A. S. C. (1977). Rudiments of numeracy. *Journal of the Royal Statistical Society, A, 140,* 277-297.

Gilks, W. R., Richardson, S., & Spiegelhalter, D. J. (Eds.). (1995). *Markov chain Monte Carlo in practice.* New York: Chapman and Hall.

Glynn, R. J., Laird, N. M., & Rubin; D. B. (1986). Selection modeling versus mixture modeling with nonignorable nonresponse. In H. Wainer (Ed.), *Drawing inferences from self-selected samples* (pp.115-142). New York: Springer-Verlag.

Goldstein, H. (1995). *Multilevel statistical models*. London: Edward Arnold; New York: Wiley.

Goodman, L. A. (1978). *Analyzing qualitative/categorical data: Log-linear models and latent structure analysis*. Cambridge, MA: Abt.

Greene, W. H. (1995). *LIMDEP Version 7.0 user's manual*. Bellport, NY: Econometric Software.

Hambleton, R. K., Swaminathan, H., & Rogers, H. J. (1991). *Fundamentals of item response theory*. Newbury Park, CA: Sage.

Harlow, L. L., Mulaik, S. A., & Steiger, J. H. (Eds.). (1997). *What if there were no significance tests?* Mahwah, NJ: Lawrence Erlbaum Associates.

Heckman, J. J., & Robb, R. (1986). Alternative methods for solving the problem of selection bias in evaluating the impact of treatments on outcomes. In H. Wainer (Ed.), *Drawing inferences from self-selected samples* (pp. 63-107). New York: Springer-Verlag.

Holland, P. W., & Rubin, D. B. (1983). On Lord's paradox. In H. Wainer & S. Messick (Eds.), *Principals of modern psychological measurement* (pp. 3-25). Hillsdale, NJ: Erlbaum.

Hosmer, D. W., & Lemeshow, S. (1989). *Applied logistic regression*. New York: Wiley.

Jöreskog, K. G., & Sörbom, D. (1979). *Advances in factor analysis and structural equation models*. Cambridge, MA: Abt.

Judd, C. M., & McClelland, G. H. (1989). *Data analysis: A model-comparison approach*. Orlando, FL: Harcourt Brace Jovanovich.

Leinhardt, G. & Leinhardt, S. (1988). Exploratory data analysis. In J. Keeves (Ed.), *Educational research, methodology, and measurement: An international handbook* (pp. 635-643). Oxford: Pergamon.

Leinhardt, S., & Wasserman, S. S. (1978) Exploratory data analysis: An introduction to selected methods. In K. F. Schuessler (Ed.), *Sociological methodology* (pp. 311-365). San Francisco: Jossey-Bass.

Little, R. J. A., & Rubin, D. B. (1987). *Statistical analysis with missing data*. New York: John Wiley.

Little, R. J. A., & Rubin, D. B. (1990). The analysis of social science data with missing values. In J. Fox & J. S. Long (Eds.), *Modern methods of data analysis* (pp. 374-409). Newbury Park, CA: Sage.

Little, R. J. A., & Schenker, N. (1995). Missing data. In G. Arminger, C. C. Clogg, & M. E. Sobel (Eds.), *Handbook of statistical modeling for the social and behavioral sciences*. New York: Plenum Press.

Little, R. J. A., & Yau, L. (in press). Statistical techniques for analyzing data from prevention trials: Treatment of no-shows using Rubin's causal model. *Psychological Methods*.

Long, J. S. (1983). *Confirmatory factor analysis*. Sage University Paper series on Quantitative Applications in the Social Sciences, 07-033. Beverly Hills, CA: Sage.

Lunneborg, C. E. (1994). *Modeling experimental and observational data*. Belmont, CA: Wadsworth.

Maddala, G. S. (1983). *Limited-dependent and qualitative variables in econometrics*. Cambridge, England: Cambridge University Press.

McCullagh, P., & Nelder, J. A. (1989). *Generalized linear models* (2nd Ed.). London: Chapman and Hall.

McCutcheon, A. L. (1987). *Latent class analysis*. Sage University Paper series on Quantitative Applications in the Social Sciences, 07-064. Newbury Park, CA: Sage.

McGaw, B. (1981). *Exploratory data analysis*. In N. L. Smith (Ed.), New techniques for evaluation (pp. 71-118). Beverly Hills, CA: Sage.

Murray, D. M., & Wolfinger, R. D. (1994). Analysis issues in the evaluation of community trials: Progress toward solutions in SAS/STAT MIXED. *Journal of Community Psychology*, (CSAP Special Issue), 140-154.

Osgood, D. W., & Smith, G. L. (1995). Applying hierarchical linear modeling to extended longitudinal evaluations: The Boys Town follow-up study. *Evaluation Review, 19*, 3-38.

Rindskopf, D. M. (1978). Secondary analysis: Using multiple analysis approaches with Head Start and Title I data. *New directions for program evaluation, 4*, 75-88.

Rindskopf, D. M. (1981). Structural equation models in analysis of nonexperimental data. In R. F. Boruch, P. M. Wortman, & D. S. Cordray (Eds.), *Reanalyzing program evaluations* (pp. 163-193). San Francisco: Jossey-Bass.

Rindskopf, D. (1986). New developments in selection modeling for quasi-experimentation. In W. M. K Trochim (Ed.), *Advances in quasi-experimental design and analysis. New directions for program evaluation, No. 31*. San Francisco: Jossey-Bass.

Rindskopf, D., & Saxe, L. S. (1998). Zero effects in substance abuse programs: Avoiding false positives and false negatives in the evaluation of community-based programs. *Evaluation Review., 22*, 78-84.

Rosenbaum, P. (1995). *Observational studies*. New York: Springer-Verlag.

Rubin, D. B. (1974). Estimating causal effects of treatments in randomized and nonrandomized studies. *Journal of Educational Psychology, 66*, 688-701.

Rubin, D. B. (1977). Assignment to treatment group on the basis of a covariate. *Journal of Educational Statistics, 2*, 1-26.

Rubin, D. B. (1978). Bayesian inference for causal effects: The role of randomization. *Annals of Statistics, 6*, 34-58.

Seltzer, M. H. (1994). Studying variation in program success: A multilevel modeling approach. *Evaluation Review, 18*, 342-361.

Singer, J. D., & Willett, J. B. (1991). Modeling the days of our lives: Using survival analysis when designing and analyzing longitudinal studies of duration and the timing of events. *Psychological Bulletin, 110*, 268-290.

Singer, J. D., & Willett, J. B. (1993). It's about time: Using discrete-time survival analysis to study duration and the timing of events. *Journal of Educational Statistics, 18*, 155-195.

Sobel, M. E. (1995). Causal inference in the social and behavioral sciences. In G. Arminger, C. C. Clogg, & M. E. Sobel (Eds.), *Handbook of statistical modeling for the social and behavioral sciences* (pp. 1-38). New York: Plenum Press.

Sommer, A. & Zeger, S. L. (1991). On estimating efficacy from clinical trials. *Statistics in Medicine, 10*, 45-52.

Sörbom, D. (1978) An alternative to the methodology of analysis of covariance. *Psychometrika, 43*, 381-396.

Spiegelhalter, D. J., Thomas, A., Best, N. G., & Gilks, W. R. (1995). *BUGS: Bayesian inference using Gibbs sampling*. Cambridge, UK: MRC Biostatistics Unit.

Thistlethwaite, D. L., & Campbell, D. T. (1960). Regression-discontinuity analysis: An alternative to the ex post facto experiment. *Journal of Educational Psychology, 51*, 309-317.

Trochim, W. M. K. (1984). *Research design for program evaluation: The regression-discontinuity approach*. Beverly Hills, CA: Sage.

Tufte, E. R. (1983). *The visual display of quantitative information*. Cheshire, CT: Graphics Press.

Tufte, E. R. (1990). *Envisioning information*. Cheshire, CT: Graphics Press.

Tukey, J. W. (1977). Some thoughts on clinical trials, especially problems of multiplicity. *Science, 198*, 679-684.

Tukey, J. W. (1977). *Exploratory data analysis*. Reading, MA: Addison-Wesley.

Wainer, Howard. (1997). *Visual revelations: Graphical tales of fate and deception from Napoleon Bonaparte to Ross Perot*. New York: Copernicus.

Chapter 10

RESEARCH SYNTHESIS

Betsy Jane Becker

INTRODUCTION

Educational productivity has been studied in a variety of ways, involving a wide range of predictors, diverse outcomes, and different levels of the educational system—for students, schools, school districts, and states. The variety of predictors of educational productivity includes such factors as class size and teacher experience (which might be impacted by educational policies), and instructional factors such as the use of tutoring and higher-order questioning in the classroom (more directly under the control of teachers themselves).

As a result of this diversity, the literature on educational productivity resists simplistic attempts to completely summarize or synthesize it into one simple result. This complexity is consistent with beliefs that different contexts, policies, and interventions likely produce accordingly diverse outcomes. The challenge for those who wish to understand educational productivity is to discern the important sources of variation in inputs, and their relationships to educational outcomes.

This chapter concerns the use of meta-analysis, or quantitative research synthesis, as a tool in facing this challenge. Meta-analysis (Glass, 1976) is a systematic approach to reviewing the empirical literature in a research domain. Meta-analysis

Advances in Educational Productivity, Volume 7, pages 193-217.
Copyright © 1998 by JAI Press Inc.
All rights of reproduction in any form reserved.
ISBN: 0-7623-0253-4

uses statistical analyses to summarize indices of study outcomes and to explore, and hopefully explain, variation in those outcomes. Modern approaches to meta-analysis (e.g., as described in *The handbook of research synthesis* edited by Cooper and Hedges, 1994) require the reviewer to focus attention on the ways that studies and their results differ, and thereby to deal with the diversity and complexity inherent in most research domains.

In this chapter, I review past attempts at quantitative synthesis of studies of educational productivity, with particular attention to the controversies and problems with such endeavors, then I provide a new perspective on the synthesis of data relevant to *systems* of variables that could be useful to understanding educational productivity.

SYNTHESES OF RESEARCH ON EDUCATIONAL PRODUCTIVITY

Much research on educational productivity is based in an economic perspective. The estimation of so-called production functions that map the contribution of various inputs, often monetary ones, to the "production" of outputs of schooling, such as student achievement, is a primary activity in such research. While some past efforts to study education production functions have involved minimal attention to the array of relevant prior research (Monk, 1992), several quantitative reviews (Greenwald, Hedges, & Laine, 1996b; Hanushek, 1989; Hedges, Laine, & Greenwald, 1994a) have focused attention and controversy on production function research. Other syntheses, not as clearly based in the economic tradition (Fraser, Walberg, Welch, & Hattie, 1987; Walberg, 1992), have aimed at pulling together an even broader set of possible predictors. As we shall see, these synthesis efforts have provided significant challenges to the researchers involved.

Hanushek's (1989) Vote-Count Summaries of Education Production Functions

Economist Eric Hanushek was among the first to systematically summarize results of education production function studies. His conclusion, that "Variations in school expenditures are not systematically related to variations in student performance" (1989, p. 45), led to calls to move away from purely expenditure-based considerations regarding school funding. It also led to further summaries (Hedges, Laine, & Greenwald, 1994a, followed by Greenwald, Hedges, & Laine, 1996b) of the empirical evidence on productivity, and to a controversy over how such summaries should be done and what has been learned from them.

Hanushek examined 187 studies (from 38 books and articles) of production functions involving seven key inputs. The inputs included overall expenditure per pupil, instructional variables (represented by teacher/pupil ratio, and teacher salary, experience, and education), and inputs related to expenses of schools (or those of school districts) such as expenditures for facilities and administration. His summaries were based on the simple and intuitively appealing procedure known as vote counting: a decision for (or against) empirical support for the importance of an input is based on counts of statistically significant results concerning that input. A typical application of the vote count is to decide that a particular input is supported if more than some proportion (often a third) of the results are significant and favorable.

For instance, Hanushek found that 16 of 65 results concerning per pupil expenditure (ppe) were significant, with 13 (or 20%) showing significance in the favorable direction. Noting that three were significant in the opposite direction, and that direction could not be ascertained for 11 others, he found "no definite indication of [ppe's] importance in determining achievement" (p. 47). Similar conclusions are drawn for the six other inputs, leading to Hanushek's general conclusion that *"There is no strong or systematic relationship between school expenditures and student performance"* (p. 47).

Hanushek considered limitations of the collected studies, and other inputs to education that he did not summarize (e.g., peer characteristics, curricular and instructional effects). Implications for policy were drawn mainly from the results of the vote counting, and he argued that policies should not be based either "principally on the basis of expenditures," (p. 49) or on the basis of surrogates for teacher and school quality (such as class size and teacher experience).

However, even a simple comparison suggests the weakness of vote counting. The 13 significant results for ppe constitute 20 percent of the total, which can be compared to the percentage expected to be positive and significant at the .05 level given *no* relationship across the populations (i.e., 2.5 percent for a two-tailed test, or at most 5% if one-tailed tests were observed). More results appear significant than would be expected if there were truly no relationship between ppe and student outcome in the population.

As it happens, vote counting has serious drawbacks as an inferential procedure. Specifically, vote counting has been found by Hedges and Olkin (1980) to have low power to detect "real" relationships (see also Bushman, 1994). Additionally, Hedges and Olkin discovered a surprising and somewhat counter-intuitive property of the vote count: the likelihood of Type II errors actually *increases* when the number of studies under review is larger. These poor statistical properties of the vote count call all decisions drawn from vote counts, but particularly decisions that favor the null hypothesis, into serious question.

Hedges, Laine, and Greenwald (1994a)

Realizing the weaknesses of vote counting, Hedges, Laine, and Greenwald (1994a) re-examined the data from Hanushek's earlier review using "more sophisticated statistical methods" (1994a, p. 5). In a first review Hedges and his colleagues applied combined significance tests and effect-magnitude analyses to Hanushek's (1989) data.

Combined Significance Tests

Combined significance tests (see, e.g., Becker, 1987, 1994) examine the simple null hypothesis that all populations show a zero parameter value—here, that all populations have a zero slope for each of Hanushek's seven inputs (that is, $H_0 : \beta_i = 0$, for studies $i = 1$ to k, tested for each input). The same directional alternative hypothesis must be tested in each study for a conventional application of combined significance tests.

Hedges and his colleagues used the combined significance test to examine both the positive alternative (H_0 versus H_1: $\beta_i > 0$), and the negative direction (H_0 versus H_2: $\beta_i < 0$). Results were expressed so that positive slopes indicated support for the importance of the input in the expected direction (i.e., more money leads to higher achievement, lower class size leads to higher achievement, etc.). One-tailed p values for tests of each directional alternative were obtained.[1]

Combined significance tests rejected the null hypothesis in favor of the alternative "at least one population shows a positive population slope" for all seven inputs for the full set of studies, and in all but one sensitivity analysis (see below). That is, there was support for the possibility of at least one studied population for which each input variable was important. For ppe, teacher experience, and teacher/pupil ratio, the results for the negative tests supported the null hypothesis, suggesting that negative population slopes were very unlikely. For the other four inputs, minimal evidence was found of possible negative slopes.

Effect Magnitudes

Hedges and his colleagues reported median regression slopes for the seven inputs. Half-standardized regression slopes were summarized for the monetary inputs ppe and teacher salary; fully standardized regression slopes were used for other inputs that could not be put on a common scale. Two main reasons precluded the application of more-typical meta-analysis summaries. First, the slope coefficients for each input were not all from the same regression model, that is, other predictors, examined with the input of interest, varied across studies. Second, different units of analysis were used across studies, which can result in different meanings, and different findings, for the "same" input variable at different levels of aggregation. (The unit-of-analysis issue is discussed further below.)

The consequence of these data characteristics is that, technically, the sample slopes under review for each input are likely not to be estimating the same population β value. While one could test the hypothesis that the results for each input variable are consistent with a single population slope (the standard meta-analytic test of homogeneity), there is little reason to expect that hypothesis to be true, given the nature of the data. Similarly, since the magnitudes of the *population* slopes might well differ to an unknown degree, interpretation of measures of spread in the observed slopes would be problematic as well.

Some might argue that even the computation of a median is suspect in light of these two features of the data. While Hedges and his colleagues did not comment on these limitations, their avoidance of analyses that address questions of consistency, and their exclusion of estimates of precision for the slopes suggest that they intend the reported median values to represent rough estimates of the central tendencies of seven populations of (possibly varying) slopes. Their analyses do not allow for tests of whether the median slopes equal zero, nor do they allow for the construction of confidence intervals. Therefore the essence of the authors' argument hinges on whether the reported median values appear large enough to have practical significance. Given the nature of the data, this is probably the best that can be expected.

Finally, Hedges and his colleagues studied three previously unexamined problematic aspects of the data (dependence, outliers, and single influential studies) via sensitivity analyses (Greenhouse & Iyengar, 1994) of special subsets of the data. Subjective assessments were made of the stability of the median values across these subsets of data.

Conclusions

Hedges, Laine, and Greenwald (1994a) argued that their findings support "a pattern of substantially positive effects for global resource inputs (PPE) and for teacher experience" (p. 11), with largely positive effects for teacher salary, administrative inputs, and facilities, and mixed results for class size. Results of the sensitivity analyses supported these conclusions. These claims, being substantially at odds with those of Hanushek (1989), led to a response from Hanushek (1994) and a rejoinder (Hedges, Laine, & Greenwald, 1994b). Not surprisingly, the authors did not modify their conclusions much in their rejoinder.

Greenwald, Hedges, and Laine (1996b)

Having found results contrary to those of Hanushek (1989) using his original data, these same authors updated and expanded the set of production function studies under review. With a more comprehensive set of studies, they conducted similar analyses, with a few further explorations. They examined historical trends via analyses of studies done since 1970, and separately examined studies using cross-sectional, quasi-longitudinal, and longitudinal designs, which offer varying degrees of control over student background. They also examined the effect of dichotomizing teacher education and teacher experience, and the use of the Test of Economic Literacy.

Conclusions

The authors again concluded that "school resources are systematically related to student achievement and that these relations are large enough to be educationally important" (Greenwald et al., 1996b, p. 384), but also argued that how money is spent in the interest of education is important as well. They contended that the expanded results showed more support for the inputs than did the prior review, with the weakest support going to class size (teacher/pupil ratio). Again a response by Hanushek (1996) did not much alter these reviewers' views (Greenwald et al., 1996a).

Fraser, Walberg, Welch, and Hattie (1987)

The review by Fraser, Walberg, Welch, and Hattie (1987) is a review of syntheses, plus a primary investigation of Walberg's educational productivity model based on national survey data. This extensive compilation of data on achievement, attitude, and other outcomes is based on nearly 300 references and thousands of original primary research studies.

Methods

The authors of this extensive review took a global approach, with the goal of assessing overall support for a number of broad domains of predictors of achievement and other school outcomes. A preliminary review of three existing models of school learning (proposed by Bloom, Carroll, and Glaser) emphasized the importance of five very general factors, specifically abililty, development (age), motivation, and two instructional factors: quantity and quality of instruction. Four additional classes of variables represented contributions of the environment, including home, classroom, peer, and media influences, as specified in prior investigations of Walberg's educational productivity model (e.g., Walberg, 1986). Variables like per pupil expenditure, class size, and school sector (i.e., public versus private school types) were not included because they were "less directly linked to student learning" (Fraser et al., 1987, p. 151).

Based on the view that "no single study...can be taken by itself as definitive" (1987, p. 151), Fraser and colleagues drew together empirical findings from a diverse set of domains, relying heavily on quantitative syntheses for their evidence. Average values of correlations or other effect-size measures were drawn from hundreds of syntheses relevant to the nine factors identified above. Additional data were drawn from analyses of large data sets such as National Assessment of Educational Progress (NAEP) and High School and Beyond.

Conclusions

To some degree, adherence to the principles of "*replication* and *generalizability*" meant that the authors of this ambitious review focused primarily on main effects—how the inputs of interest work for most students, across different school subjects, in many grades, and so on. This approach may have reduced the applicability of the empirically-derived model to any specific child, whose individual characteristics render him or her different from the average child. Fraser and his co-authors anticipated this critique by noting "the results are surprisingly robust. The more powerful factors appear to benefit all students in all conditions; but some students appear to benefit somewhat more than others under some conditions" (1987, p. 153). The idea that some students might not benefit at all, or in less likelihood be harmed, was not addressed.

In several chapters of this monograph, the authors examined data from research syntheses. One chapter is devoted to science outcomes, while the most extensive survey covers 134 meta-analyses on a variety of kinds of achievement and 92 meta-analyses of the prediction of attitudes. They then discussed the largest average effects, comparing each to ability and background effects and assessing their likely impact on school outcomes.

As the authors noted, many details of the original research and meta-analyses were omitted in the attempt to synthesize so much. However, they felt confident based on this sweeping overview that suggested improvements "seem likely to increase teaching effectiveness and educational productivity" (Fraser et al., 1987, p. 164).

Other Reviews

Clearly, in some broad sense, any review of school-related predictor variables and school outcomes is relevant to the study of educational productivity. This could include reviews of school-based educational interventions (e.g., within-class grouping, studied by Lou et al., 1996), general instructional methods (e.g., computerized instruction, reviewed by Fletcher-Flinn & Gravatt, 1995), or correlates of school performance (e.g., Pajares' 1996 review of self-efficacy), and even reviews of group differences such as gender effects, if these lead to systematic differences in school outcomes (e.g., Linn, 1992). However, since most reviews of these other domains have not explicitly focused on development of a comprehensive set of factors to explain school outcomes, they are not considered here.

IMPORTANT MODERATOR VARIABLES IN RESEARCH ON EDUCATIONAL PRODUCTIVITY

A key task in a successful research synthesis is to explore the study features that might relate to differences in study results. Such explanatory variables, or predictors of study outcomes, are called moderator variables. Moderator variables in meta-analysis may reflect differences in subject populations (e.g., age or gender of subjects), in study design (e.g., method of sample selection, type of measure used) or more substantively interesting features such as duration or nature of treatment (if one is present). In this section I discuss several critical moderator variables in studies of educational productivity, and why they should be of particular interest to reviewers.

As mentioned above, studies of educational productivity are diverse indeed. While on the one hand this diversity clearly leads to questions about the similarity of results across studies within a meta-analysis, it need not pose a problem to the synthesis. In fact, generally speaking, the greater the diversity in the kinds of studies that are represented, the more likely it is that the reviewer will be able to assess the generalizability of results across the conditions studied, and to generalize across a broad range of conditions.

Broad generalizations are most justifiable when studies vary greatly in design, in the cases or units studied, in settings, and even in how inputs and outcomes of schooling are measured, *and* those variations appear unrelated to study results. Cook has referred to such study features as heterogeneous irrelevancies, noting, "The greater the heterogeneity in the array of persons, settings, measures, and the like across which an effect is replicated, the greater is the confidence that the effect is universally generalizable, even to unexamined contexts" (Cook, 1990, p. 26).

In the case of educational productivity research there are a few exceptions to this rule. Specifically, critical concerns include differences in the unit of analysis across studies (i.e., level of aggregation of data), and in the composition of the model(s) that are investigated (the predictors and outcomes used). I deal with those first, then briefly discuss several other important variables.

Critical Moderators

Different Units of Analysis Imply Different Slopes

Studies of educational productivity exist at all levels of aggregation. Seventy percent of the samples in Greenwald and colleagues' (1996b) review reported results at the student level, while less than 20 percent were based on district-level data. Data from students, schools (e.g., Michelson, 1972), school districts (e.g., Kiesling, 1967), and states (Behrendt, Eisenach, & Johnson, 1986) involve variables that have different meanings and that can be manipulated to a greater or lesser degree by policymakers. Some variables that exist at one level do not have analogues at lower levels of aggregation, for instance, measures of spread on student background characteristics at the classroom level have no parallel at the student level if one test was given to each student within each class. Similarly, per pupil expenditure is typically computed at the school or district level, and in many cases it is unlikely that the exact expenditure for any given pupil could be computed at all. Both output and input variables, and the slopes representing the importance of the inputs, may have different interpretations at different levels of aggregation even when those variables exist at all levels.

As the level of aggregation rises, so also does the distance between the persons who can change or control relevant variables and the original populations of school children, who produce the educational outcomes. State increases in per pupil spending levels may never be experienced by the average student (or *any* student), as funds are directed to students in particular locations or grades, spent on special programs, or used for administrative expenses at the state level. Vast disparities in funding levels between states, among districts, and among schools within districts (Odden & Clune, 1995) raise questions about the meaning of aggregate funding measures for any one case within each level of aggregation.

An extensive and long-standing literature in sociology explores the aggregation problem (e.g., Goodman, 1953; Robinson, 1950). A variety of aspects of aggregation has been studied, including the roles of within and between-groups variation and measurement issues   (see, e.g., Burstein & Linn, 1982; Sirotnik & Burstein, 1985). While some have argued that aggregated data may be more accurate because measurement error at lower levels is reduced (Loeb & Bound, 1996, p. 661), others (Hanushek, Rivkin, & Taylor, 1996) have found that aggregation can exacerbate the problem of omitted-variable bias (e.g., via omission of appropriate controls for student background). A general conclusion of this literature is that, in general, it is not safe to use data at higher levels of aggregation to make inferences about lower-level units.

Multivariate Nature of the Data Leads to Different Slopes

The multivariate nature of models for productivity contributes several sources of variation to the observed regression slopes. Studies are multivariate in both their inputs and outputs, and as in many fields, it is unusual to find a literal replication of any educational productivity study. Each researcher adds a particular twist or modification of what was done before to make his or her study unique.

One consequence of the use of multiple variables is that, across studies, many different sets of inputs have been considered. Thus, slopes for the same input may not estimate the same parameter because they represent different partial relationships. Early on, smaller sets of predictors were considered adequate. Kiesling wrote, "Formal education is only one of three basic influences upon a person's educational progress, the other two being native intelligence and motivation toward learning" (1967, p. 357). Since that time many more kinds of inputs have come to be recognized as potentially important; Summers and Wolfe (1977) included 20 predictor variables and nine additional interaction terms among them.

Even when researchers start with the same set of variables, they may transform them (the log transformation is popular, as in Michelson, 1972), combine them to create interactions (e.g., Tuckman, 1971), or aggregate them in a different way so as to create a different set of variables for analysis. The addition of new and *relevant* predictors to a model can lead to different findings for an input that was previously modeled[2] because the interpretation of each regression slope is made in the context of the other variables "held constant." To the extent that the predictor of interest is collinear with other predictors, its slope and the precision (and significance) of that slope can change simply because other variables are added to the model (Fox, 1991).

Many school-resource inputs are structurally related, sometimes producing multicollinearity which may not have been investigated or reported. For instance, total per pupil expenditures are largely a function of teacher salaries, but both variables may be included in a single regression model. Therefore, within a study, slopes for different inputs are rarely totally independent. This fact is further complicated when across studies, some studies contribute more-or-less independent slopes and others contribute slopes that are heavily affected by multicollinearity.

Differences in Sample Composition

A number of facets of sample composition are relevant in educational productivity research, including representativeness. While a surprising number of studies has been based on student-level data from nationally representative probability surveys such as Project TALENT (e.g, Kenny, 1982; Perl, 1973) and NELS:88 (Brewer, 1996), studies not based on large surveys can have very restricted samples (e.g., Winkler, 1975, studied students from one California school district). It is also quite common for studies to be based on data from schools (e.g., Michelson, 1972), and school districts, typically studied within a single state (e.g., Strauss & Sawyer, 1986).

Differential Selection

Wainer (1993) showed the effects of differential selection via a very simple pair of graphs, using data on within-state average ppe and two outcomes: state rank on mean Scholastic Aptitude Test (SAT) scores and state rank on scores from the National Assessment of Educational Progress (NAEP). The data of SAT rank and state ppe values had been used in a *Wall Street Journal* article to argue that money spent did not lead to better rankings on SAT scores. Indeed, Wainer's graph shows that state rank on SAT increases (worsens) by about 5.5 positions with every additional $1000 spent.

However, the SAT is well-known to be taken by a very nonrandom subset of students—college-bound students who tend to be high achievers. Also the proportion of students who take the SAT varies across states. The NAEP, however, is administered to what are designed to be representative samples of examinees in each state. Notably, the slope relating mean ppe to NAEP rankings shows that a state's

NAEP rank is expected to improve by about two positions for every added $1000 spent, across the states.

Is selectivity a problem in the literature on productivity? The answer is "yes," to the extent that outcomes associated with different levels of self-selection are used across studies, or that differential selection occurs even with the same outcomes. Specifically, Behrendt and colleagues (1986) used the SAT (it should be noted that these authors focused on selectivity bias), whereas others have used outcomes measured on data from representative samples, including the NAEP (Wenglinsky, 1997).

Range Restriction

A second aspect of sample composition with consequences similar to those differential selection is range restriction. Magnitudes of observed bivariate correlations are reduced when values for cases in the sample are restricted on either variable. Samples of large schools (e.g., Burkhead, 1967), urban and inner city schools (e.g., Katzman, 1971; Murnane, 1975), or subjects who have been part of a specific social or educational intervention (Maynard & Crawford, 1976) may show weaker relationships if the ranges of values of input or outcome variables, or other variables correlated to those, are restricted.

Range restriction has received much attention in the literature on validity generalization (Hunter & Schmidt, 1990b), a domain which frequently uses methods for meta-analysis. Range restriction is of critical importance in understanding the use of tests for predicting job performance when hiring decisions are made on the basis of test performance. The observed validity of the selection test—its correlation with some job outcome—depends directly on the degree of range restriction. When different selection criteria are used across settings or job types, corresponding differences in observed validities may be observed even when the true validities (population correlations) are equal. Similar effects may be seen in the literature on educational productivity.

If explicit selection criteria are used it is easy to get information on the degree of range restriction from standard deviations for the selected and unselected groups on the selection variable, and simple corrections can be made to adjust for differential sample selection (e.g., Hunter & Schmidt, 1990b). However, when subjects are selected on the basis of variables that are (at best) correlated with what is measured, and which may or may not be measured themselves, it may be impossible to determine the degree of range restriction that applies to the measured predictors. This is likely to be the case in many studies of educational productivity. Without such information, corrections cannot be made, and some unknown amount of observed variation in strength of relationships in the sample data (e.g., the slopes) may be caused by differential range restriction.

Measurement Issues

The use of different types of measures, and even measures of slightly different conceptualizations of inputs and outputs can often lend generalizability to the results of a synthesis. However, several issues related to measurement are sources of unwanted variation, and the meta-analyst who desires to understand variation in results as a representation of true differences in the importance of various inputs should try to minimize, correct for, or at least investigate, these issues.

Inflation

Currency inflation, and its companion deflation, has important consequences in studies of monetary inputs to education. Variation over time in value of expenditures means that inputs of $100 at two time points can have very different consequences simply because the value of money has changed. In reviews of productivity research the effect is seen in the interpretations that are made of slope coefficients for monetary inputs. In the United States, inflation has been the historical trend, and it is typical to apply corrections to study results for changes due to inflation in syntheses where magnitudes of slope coefficients are of interest (e.g., Greenwald et al., 1996b).

Measurement Error

There is measurement error in virtually all recordings of information about human characteristics and performances. However, the consequences of this error for statistical inference are frequently ignored within the confines of a single study. Occasionally, but not always, the reader can find reports of the reliabilities of the instruments used to measure outcomes and predictors in educational research. It is much more unusual for those reliabilities to be incorporated into data analyses, though an extensive literature exists on how to accomplish such adjustments (e.g., Fuller & Hidiroglou, 1978; Gleser, 1992), including in meta-analysis (again see Hunter & Schmidt, 1990b).

The literature on errors-in-variables regression indicates that slopes and their standard errors can be affected by measurement error, which reduces the degree of association that can be observed. To the extent that different levels of measurement error exist across studies, this will be another source of variation in the observed slopes. If corrections can be applied this variation can be explained in a research synthesis. However, information about reliability tends not to appear in studies of educational production functions (e.g., Hanushek, 1972; Levin, 1976; Murnane, 1975; Sebold & Dato, 1981), thus it is unlikely the meta-analyst will be able to fully account for differential attenuation.

Dichotomization

The last issue concerns dichotomization, or more generally creating categorical variables from continuous data. Hunter and Schmidt (1990a) have shown that dichotomization leads to two distortions in syntheses of correlations: average values tend to be lower, and the apparent variability of the correlations is increased. This issue is complicated by the fact that the extent of distortion depends on where the distribution of scores is split. For the case of dichotomies, Hunter and Schmidt (1990a) provide corrections that can be applied. When corrections for this feature are not made, the meta-analyst may again be unable to fully account for observed variation across studies.

A MULTIVARIATE SYNTHESIS APPROACH

The syntheses reviewed above reveal much about research on educational productivity. However, because of a variety of limitations of the syntheses, primarily in the data available for synthesis, many questions remain about how to effect the most positive educational outcomes. In this final section of the chapter, I illustrate a multivariate, model-based approach to meta-analysis (Becker, 1992, 1995; Becker & Schram, 1994). Given appropriate primary-study data, this approach allows the reviewer to examine questions that cannot be addressed with any of the methods described above and applied to research on educational productivity to date.

Direct and Indirect Effects

One important feature of this multivariate approach is that it allows the reviewer to examine systems of variables, not just simple bivariate predictor-outcome relationships. Consequently, not only can the reviewer look for direct impacts of each potential predictor on an outcome, but he or she can also explore the possibility that a variable has indirect influence through the mediating effects of one or more intermediate variables.

An interesting example of just such a finding comes from a meta-analysis by Whiteside and Becker (1996) of the literature on child outcomes in divorcing families. Their investigation was based on a set of direct and indirect relations specified on the basis of prior theory and other research on divorce and family dynamics. Previous narrative research reviews in this domain had questioned the influence of the extent of the father's visits on child outcomes, based on the finding that the direct relationship between these two variables was not particularly strong. In a model-based analysis of studies of very young children, Whiteside and Becker examined the possibility of an indirect connection between these two variables, mediated by the quality of the father-child relationship.

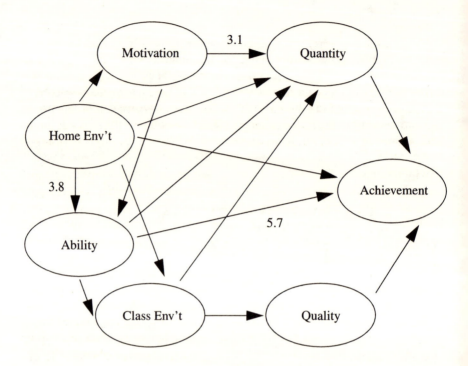

Figure 10.1.   Reduced Version of Walberg's Educational Productivity Model

Their analysis confirmed a weak direct link between father visits and child outcome. However, they also discovered strong (and significant) connections
between father visits and father-child relationship, and between father-child
relations and positive child outcomes. Thus using this multivariate approach to
meta-analysis, Whiteside and Becker could make specific recommendations
about the role of father visitation, one of the few variables in the constellation
of available predictors that can be impacted by the courts in child custody decisions.

Figure 10.1 shows a reduced version of Walberg's educational productivity
model with four direct effects. This model is examined below in a synthesis of four
studies. The direct influence of home environment on student learning was investigated here, a link that does not appear in either model tested in Reynolds and
Walberg's (1991, 1992) primary research. Peer influences, which Reynolds and
Walberg viewed as a mediator of the effects of the home, do not appear in this
model. Similarly, a factor representing media influences served as a mediator for
ability, motivation, and home environment factors in full versions of the model,
but was omitted in this example. This analysis cannot be interpreted as a test of

Walberg's productivity model, even for this small data set, but it will illustrate how both types of effects can be studied.

Questions Addressed by the Multivariate Approach

The multivariate approach to synthesis is based in large part on the estimation of a correlation matrix from the series of studies. The interpretation of this matrix depends on whether the reviewer adopts a fixed- or random-effects model for the data. Choice of a model depends on the assumptions the reviewer is willing to make.

With the simplest fixed-effects approach, one assumes that all observed variation in the correlation matrices can be attributed to sampling error. This is equivalent to assuming that a single population correlation matrix underlies every relationship examined in every study in the review. This model is typically not expected to hold, because studies vary in a variety of ways, many of which may be of interest. However, it is easy to test whether the fixed-effects model holds using homogeneity-test methods proposed by Becker (1992). If the hypothesis of homogeneity (consistency) is accepted, the average correlation matrix can be viewed as an estimate of the common population matrix, shared by all studies.

The simplest random-effects model assumes that all studies are sampled from a population of studies, and that the population correlations for each relationship may differ across studies. When one rejects the hypothesis of homogeneity (the fixed-effects model), it often makes sense to estimate a measure of spread among the population correlations for each relationship.

Further, this model assumes that differences in population correlations do not relate to specific study features such as type of sample, age of students, or type of outcome measure. While it may make sense to allow the measures of the predictor and outcome constructs to differ, the reviewer should believe that the constructs (underlying variables) are the same. There may still be debates over whether those constructs are, indeed, the same (e.g., are math and reading achievement "the same" because both represent achievement), but they must be similar enough so that an average population correlation, averaged across the variety of constructs included, is interpretable. In other meta-analytic applications the variables to be combined are required to be linearly equatable, and that is a reasonable requirement to place on the constructs underlying the correlation matrix as well.

The real state of affairs is probably somewhere between these two simplest models—in other words, a mixed model is likely best for most data. Mixed models apply when some explanatory variables differentiate the study results, but they do not explain all of the observed between-studies variation in the data.

After a common or average correlation matrix is estimated from the studies, that matrix can be used to estimate and examine linear models that specify both direct and indirect relationships of predictor variables with outcomes of interest.

Data Requirements

Only the sample sizes and zero-order correlations among the variables in the primary research are needed to apply this multivariate synthesis approach. However, even these simple data have generally not been reported in the literature on economic production functions. In fact, only six of Hanushek's (1989) studies provide correlation matrices (Bowles, 1970; Cohn, 1968; Dynarski, 1987; Katzman, 1971; Raymond, 1968; Ribich & Murphy, 1975). While initial model-based analyses have explored the results of those studies (Becker & Kamata, 1994), the studies use different units of analysis. Variation in the results of these six studies may not be attributable simply to true differences in the magnitudes of interrelationship among the variables. Therefore another data set, described below, was selected to illustrate this approach.

Details of the multivariate meta-analysis estimation and tests are described more fully in Becker (1992, 1995) and Becker and Schram (1994). In short, the procedures are based on generalized least squares (GLS) estimation methods which incorporate information on variation and covariation among the correlations to obtain weighted average correlations and estimates of their uncertainty. In the case of random-effects modeling, between-studies variation is estimated and incorporated. Then the average correlations (and their variance-covariance matrix) are used to compute standardized regression equations representing theoretical models of direct and indirect relationships of inputs and outcome.

Data

The data for this illustration are from four studies of Walberg's educational productivity model (Ibe, 1994; Reynolds & Walberg, 1991, 1992; Sares, 1992). This is neither an exhaustive set of studies, nor an identifiable subset of the literature on this productivity model. However, since correlation matrices among variables representing six inputs in Walberg's model were available in these studies, they constituted a convenient subset for this example.

Data from the studies are shown in Table 10.1. Correlations are available for the factors ability, motivation, quality and quantity of instruction, home environment, classroom environment, and the outcome of student achievement. Each study used student-level data, from large nationally representative data sets. Ibe (1994) studied data originally collected in 1972 and 1979 from high schoolers in the National Longitudinal Study (NLS), Reynolds and Walberg (1991, 1992) used samples from grades eight and 11, respectively, from the Longitudinal Study of American Youth, and Sares (1992) studied U.S. students from the Second International Mathematics Study.

One notable feature of these studies is that each is based on a relatively large sample size. The analyses below are weighted by the level of precision of each study's results, and each data point included in the analysis is quite precise, in

Table 10.1. Correlation Matrices from Four
Studies of Walberg's Productivity Model

	Ability	Motiv'n	Quality	Quantity	Home	Class	Achv't
Sares (1992) $n = 1539$							
Ability	1	.012	.130	.103	.336	.253	.784
Motiv'n		1	.007	.131	.005	.121	.007
Quality			1	.045	.077	.001	.086
Quantity				1	.089	.024	.124
Home					1	.100	.330
Class						1	.285
Ibe (1994) $n = 7935$							
Ability	1	.014	.020	.142	.051	−.012	.231
Motiv'n		1	.009	.008	.020	−.028	−.011
Quality			1	.019	−.005	.027	.066
Quantity				1	.026	.098	.231
Home					1	.060	.023
Class						1	.017
Reynolds & Walberg (1991) $n = 3116$							
Ability	1	.241	.255	.457	.542	.390	.689
Motiv'n		1	.128	.446	.129	.167	.279
Quality			1	.100	.272	.443	.295
Quantity				1	.343	.389	.529
Home					1	.439	.477
Class						1	.384
Reynolds & Walberg (1992) $n = 2535$							
Ability	1	.080	.162	.285	.535	.412	.736
Motiv'n		1	−.002	.660	.221	.151	.119
Quality			1	.179	.297	.460	.173
Quantity				1	.395	.262	.297
Home					1	.546	.479
Class						1	.335

an absolute sense. It is likely that the results will appear inconsistent or heterogeneous simply because the standard errors of each correlation are quite small. Differences between correlations as small as .06 can reach significance for samples of 1500 cases, and all four samples are larger than that.

While all four studies provided 21 unique correlations among the seven listed factors, measures of most factors varied slightly across studies. For instance, both studies by Reynolds and Walberg considered science achievement as the outcome, while the other studies involved math (Sares) or math and verbal skills combined (Ibe). Similarly, while hours of homework reported by the student were a component of the quantity-of-instruction measure in all studies, Reynolds and Walberg (1991) added an attendance indicator, and Reynolds and Walberg (1992) added both a dummy variable representing cutting class and a parental report of homework time. There is face validity for the use of each of these measures as indicators of the named factors in Walberg's model. However, whether the measures actually provide similar results is an empirical question.

Finally, other variables not common to all four samples have been omitted to create a simple complete data set. In practice, few series of studies in any domain use identical sets of variables, but these synthesis techniques can be adapted for incomplete data (Becker & Schram, 1994).

Results

The values above the diagonal in the first matrix in Table 10.2 show the average fixed-effects correlations for the $p^* = 21$ relationships among the seven factors in the reduced Walberg model. Under the null hypothesis that all studies share a common population correlation matrix, the test of homogeneity, denoted Q_T, is a chi-square test. Here the test is significant, with $Q_T = 8038.3$ ($p < .001$, $df = p^*(k-1) = 21 \times 3 = 63$), indicating, as was expected, that all studies do not appear to share a common population correlation matrix. The large within-study sample sizes for these four studies lead to very high levels of precision for the raw data and for the fixed-effects means. (The effective sample size for the analysis is 15,125.) All means in this matrix appear to differ from zero under the fixed-effects model, according to individual z tests of the hypothesis H_0: $\rho = 0$ for each relationship (at the $\alpha = .001$ level).

Below the diagonal in Table 10.2 are variance components for each of the 21 relationships, presented in the metric of the Fisher z transformation, which is used for computations. The z metric is used because no transformation exists to return the variances to the correlation metric. These variances can be added to the within-study sampling variances to incorporate uncertainty due to between-studies differences into further analyses. Also, the variances indicate spread in the distributions of the population correlations, assuming the random-effects model holds. While the degrees of freedom for hypothesis tests do not change under random-effects assumptions, most standard errors will increase when the population variances are incorporated, thus lowering the values of test statistics such as the tests of H_0: $\rho = 0$ (mentioned above) and the z tests of slopes (as in Table 10.3 below).

Because the overall test of homogeneity indicated that the model of a common population matrix was not appropriate, the random-effects model was adopted. Average correlations were obtained under this model, which views the uncertainty in each study result to be composed of within-study sampling error plus random population variation. When population variation is large relative to sampling variation, as is the case in these data, studies are weighted more equally under the random-effects model. The second matrix in Table 10.2 shows the means under the random-effects model. Some values differ from the fixed-effects means because they are not as heavily influenced by the results from Ibe (the largest study). Also over half of the correlations no longer differ from zero, due to the added population variation. (The $\alpha = .02$ significance

Table 10.2. Correlations (Above Diagonal) and
Variance Components[1] for Walberg's Productivity Model

	Ability	Motiv'n	Quality	Quantity	Home	Class	Achv't
Fixed-effects Model[2]							
Ability	1	.073	.105	.232	.280	.176	.510
Motiv'n	.013	1	.032	.244	.075	.058	.074
Quality	.014	.004	1	.065	.114	.193	.135
Quantity	.032	.144	.005	1	.165	.182	.299
Home	.104	.010	.030	.045	1	.237	.237
Class	.065	.013	.078	.028	.083	1	.179
Achv't	.194	.021	.013	.035	.086	.047	1
Random-effects Model[3]							
Ability	1	.088	<u>.143</u>	**.253**	<u>.381</u>	.267	**.647**
Motiv'n	.013	1	.036	.339	.095	.103	.100
Quality	.014	.004	1	.086	.163	.245	**.156**
Quantity	.032	.144	.005	1	.219	<u>.198</u>	**.304**
Home	.104	.010	.030	.045	1	.301	<u>.338</u>
Class	.065	.013	.078	.028	.083	1	<u>.259</u>
Achv' t	.194	.021	.013	.035	.086	.047	1

Notes: 1. Variance components are presented in the metric of the Fisher z transformation. Relative comparisons among these values for the 21 relationships are valid.
2. All fixed-effects mean correlations differ significantly from zero at $\alpha = .001$.
3. Average random-effects mean correlations shown in boldface differ significantly from zero at $\alpha = .001$. Values that are either underlined or boldfaced differ from zero at $\alpha = .02$.

level was used to protect against excess type I error levels due to the many tests within each matrix.)

The random-effects averages, and their variance-covariance matrix (not shown here) were used to estimate a series of regression models representing the paths in Figure 10.1. The standardized regression results are shown in Table 10.3, and the three significant paths are shown in bold in Figure 10.1.

Each potential predictor is viewed as an outcome of the predictors that lead to it. Achievement is the final direct outcome of four inputs. Standardized slope coefficients for each path (b values), standard errors, and z tests of the significance of each slope are provided. Ability has a direct impact on achievement in spite of its large variance component, and home environment has an indirect role through its association with ability. Motivation is a significant predictor of quantity of education (primarily time spent on homework), and quantity is positively, but not significantly related to outcome. Other variables do not appear to have significant effects on achievement under the random-effects model.[3]

Of interest for interpretation and policymaking is the possibility of detecting indirect relationships, such as that for home environment. In the absence of the home-ability connection, home factors would not be viewed as important to school outcomes, because the direct impact of home in the final regression

Table 10.3. Random-Effects Standardized Regression
Coefficients and Standard Errors for the Walberg Model

Outcome	Predictor	Standardized slope (b)	Standard error of b	z	p
Ability					
	Motivation	0.052	0.063	0.83	.40
	Home	0.376	0.163	2.30	.02
Motivation					
	Home	0.095	0.050	1.88	.06
Class					
	Ability	0.178	0.199	0.89	.37
	Home	0.234	0.167	1.40	.16
Quality					
	Class	0.245	0.140	1.75	.08
Quantity					
	Ability	0.163	0.121	1.35	.18
	Motivation	0.306	0.097	3.15	.0016
	Home	0.100	0.210	0.48	.63
	Class	0.093	0.157	0.59	.56
Achievement					
	Ability	0.574	0.122	4.72	*
	Quality	0.050	0.101	0.49	.62
	Quantity	0.137	0.188	0.73	.47
	Home	0.081	0.273	0.30	.77

Note: The z value is computed as $b/SE(b)$. Asterisk indicates observed probability less than .0001.

model is small (less than a tenth of a standard deviation of change in outcome is predicted), and nonsignificant.

In a real application involving more studies, next steps would include exploration of different models (other connections among inputs and outcome), between-studies differences, or the effect of eliminating outliers on results. In this data set the correlations from Ibe might be omitted, since those results appear weakest across all relationships.

SUMMARY

Meta-analysis has helped, at a minimum, to bring to the fore some issues that have stymied researchers and policymakers in the area of educational productivity, and to focus attention on the design issues and other factors that may help explain why resources seem to matter in some cases but not in others. Given the complexity of this field, it is likely that the study of educational productivity will continue to present challenges to researchers and reviewers alike.

When studies of educational productivity are synthesized, several issues must be addressed. Both readers (consumers) and producers of future syntheses of this literature should attend to these issues. First, reviews of this literature must acknowledge the presence of aggregation bias if studies with differing units of analysis are combined. This varying design aspect of educational productivity studies seriously limits the possibility of creating a broadly inclusive set of comparable effect-magnitude measures that are amenable to quantitative analysis. The problem can be avoided by using less informative analyses such as tests of combined significance, or by examining smaller subsets of studies all sharing the same units of analysis.

In avoiding the first problem, the reviewer must take care not to cause other problems, such as creating subsets of studies that are too limited to warrant quantitative synthesis. These subsets can be limited in number (that is, containing only a few studies), or in kind, whereby the studies deemed "similar enough" to combine are so similar that any generalizations based on them will not be well-founded. In the educational productivity domain, the former problem is much more likely to occur than the latter. Existing diversity of inputs, outcomes, types of subjects, and the like seems sufficient to suggest that finding even two identical studies is unlikely.

Second, as in all research reviews, syntheses of educational productivity research must focus attention on those between-studies differences that are theoretically important to study outcomes. If a researcher has argued, say, that differences in class size impact student outcome only when classes are small, then knowing the typical class size in each study will be critical to addressing the importance of class size as an input. While it is also important to examine between-studies differences in study design and method, it is imperative to consider the role of theoretically (substantively) important variables, because they form the basis of our understanding of how the phenomenon of interest—here, educational productivity—works in the real world (not in a designed empirical study of the real world).

It is hoped that in the future, even more informative quantitative research syntheses of educational productivity will be possible, enabling reviewers not only to look broadly at whether resources are important at all, and what kinds of inputs matter, but to explore how available resources can be most effectively put to the cause of improving school outcomes.

NOTES

1. The data for the test of H_1 were observed directional significance values p_{ji}, for the jth input, across all k studies that had presented slopes for the jth input. The p value for the test of H_2 is $(1-p_{ji})$, where p_{ji} is the value for the test of H_1. Fisher's (1932) test was used; for the positive test of the jth input it is $-2 \Sigma_i \log (p_{ji})$, and for the negative test it is $-2 \Sigma_i \log (1-p_{ji})$.

2. This holds true within studies because of multicollinearity, and between studies of the same population. If two studies differ in both the set of inputs included in their regression models *and* the populations under study, between-studies differences in slopes cannot be attributed to multicollinearity alone.

3.   All paths in the model were significant under fixed-effects assumptions. Between-studies differences may explain some of the variation in results, but with only four studies no tests of moderators were done.

REFERENCES

Becker, B.J. (1987). Applying tests of combined significance in meta-analysis. *Psychological Bulletin, 102*, 164-171.

Becker, B.J. (1992). Using results from replicated studies to estimate linear models. *Journal of Educational Statistics, 17*, 341-362.

Becker, B.J. (1994). Tests of combined significance. In H.M. Cooper & L.V. Hedges (Eds.), *The handbook of research synthesis*. New York: Russell Sage.

Becker, B.J. (1995). Corrections to "Using results from replicated studies to estimate linear models." *Journal of Educational Statistics, 20*, 100-102.

Becker, B.J., & Kamata, A. (1994, April). *Another look at the relationship between school expenditures and students' achievement.* Paper presented at the annual meeting of the American Educational Research Association, New Orleans, LA.

Becker, B.J., & Schram, C.M. (1994). Examining explanatory models through research synthesis. In H.M. Cooper & L.V. Hedges (Eds.), *The handbook of research synthesis*. New York: Russell Sage Foundation.

Behrendt, A., Eisenach, J., & Johnson, W.R. (1986). Selectivity bias and the determinants of SAT scores. *Economics of Education Review, 5*, 363-371.

Bowles, S. (1970). Toward an educational production function. In W.L. Hansen (Ed.), *Education, income and human capital* (pp. 11-60). New York: National Bureau of Economic Research.

Brewer, D.J. (1996). Does more school district administration lower educational productivity? Some evidence on the "administrative blob" in New York public schools. *Economics of Education Review, 15*(2), 111-124.

Burkhead, J. (1967). *Input-output in large city high schools.* Syracuse, NY: Syracuse University Press.

Burstein, L., & Linn, R.L. (1982). *Analysis of educational effects from a multilevel perspective: Disentangling between- and within-class relationships in mathematics performance.* University of California, Los Angeles. Center for the Study of Evaluation. (ERIC Document Reproduction Service Number ED 228 311).

Bushman, B.J. (1994). Vote-counting. In H.M. Cooper & L.V. Hedges (Eds.), *The handbook of research synthesis*. New York: Russell Sage.

Cohn, E. (1968). Economies of scale in Iowa high school operations. *Journal of Human Resources, 3*, 422-434.

Cook, T.D. (1990). The generalization of causal connections: Multiple theories in search of clear practice. In L. Sechrest, E. Perrin, & J. Bunker (Eds.), *Research methodology: Strengthening causal interpretations of nonexperimental data*. Washington, DC: U.S. Department of Health and Human Services.

Cooper, H.M., & Hedges, L.V. (Eds.). (1994). *The handbook of research synthesis.* New York: Russell Sage Foundation.

Dynarski, M. (1987). The Scholastic Aptitude Test: Participation and performance. *Economics of Education Review, 6*, 263-274.

Fisher, R.A. (1932). *Statistical methods for research workers* (4th ed.). London: Oliver & Boyd.

Fletcher-Flinn, C.M., & Gravatt, B. (1995). The efficacy of computer assisted instruction (CAI): A meta-analysis. *Journal of Educational Computing Research, 12*(3), 219-241.

Fox, J. (1991). Regression diagnostics. *Quantitative applications in the social sciences.* University Paper Series, 79. Beverly Hills, CA: Sage.

Fraser, B.J., Walberg, H.J., Welch, W.W, & Hattie, J.A. (1987). Syntheses of educational productivity research. *International Journal of Educational Research, 11*, 145-252.

Fuller, W.A., & Hidiroglou, M.A. (1978). Regression estimation after correcting for attenuation. *Journal of the American Statistical Association, 73*, 99-104.

Glass, G.V (1976). Primary, secondary, and meta-analysis of research. *Educational Researcher, 5*, 3-8.

Gleser, L.J. (1992). The importance of assessing measurement reliability in multivariate regression. *Journal of the American Statistical Association, 87*, 969-707.

Goodman, L. (1953). Ecological regressions and behavior of individuals. *American Sociological Review, 18*, 663-664.

Greenhouse, J.B., & Iyengar, S. (1994). Sensitivity analysis and diagnostics. In H.M. Cooper & L.V. Hedges (Eds.), *The handbook of research synthesis*. New York: Russell Sage.

Greenwald, R., Hedges, L.V., & Laine, R.D. (1996a). Interpreting research on school resources and student-achievement: A rejoinder to Hanushek. *Review of Educational Research, 66*(3), 411-416.

Greenwald, R., Hedges, L.V., & Laine, R.D. (1996b). The effect of school resources on student-achievement. *Review of Educational Research, 66*(3), 361-396.

Hanushek, E.A. (1972). *Education and race: An analysis of the educational production process*. Cambridge, MA: Heath-Lexington.

Hanushek, E.A. (1989). The impact of differential expenditures on school performance. *Educational Researcher, 18*(4), 45-65.

Hanushek, E.A. (1994). Money might matter somewhere: A response to Hedges, Laine, and Greenwald. *Educational Researcher, 23*(4), 5-8.

Hanushek, E.A. (1996). A more complete picture of school resource policies. *Review of Educational Research, 66*(3), 397-409.

Hanushek, E.A., Rivkin, S.G., & Taylor, L.L. (1996). Aggregation and the estimated effects of school resources. *Review of Economics and Statistics, 78*(4), 611-627.

Hedges, L.V., Laine, R.D., & Greenwald, R. (1994a). Does money matter? A meta-analysis of studies of the effects of differential school inputs on stduent outcomes. *Educational Researcher, 23*(4), 5-14.

Hedges, L.V., Laine, R.D., & Greenwald, R. (1994b). Money does matter somewhere: A reply to Hanushek. *Educational Researcher, 23*(4), 9-10.

Hedges, L.V., & Olkin, I. (1980). Vote counting methods in research synthesis. *Psychological Bulletin, 88*, 359-369.

Hunter, J. E., & Schmidt, F. L. (1990a). Dichotomization of continuous variables: The implications for meta-analysis. *Journal of Applied Psychology, 75*(3), 334-349.

Hunter, J. E., & Schmidt, F. L. (1990b). *Methods of meta-analysis: Correcting error and bias in research findings*. Newbury Park, CA: Sage Publications.

*Ibe, R.E. (1994). The enduring effects of productivity factors on eighth grade students' mathematics outcome. Paper presented at the annual meeting of the American Educational Research Association, New Orleans, LA. (ERIC Document Reproduction Service Number ED 374 968.)

Katzman, M. (1971). *Political economy of urban schools*. Cambridge, MA: Harvard University Press.

Kenny, L.W. (1982). Economies of scale in schooling. *Economics of Education Review, 2*, 1-24.

Kiesling, H.J. (1967). Measuring a local school government: A study of school districts in New York state. *Review of Economics and Statistics, 49*, 356-367.

Levin, H.M. (1976). Economic efficiency and educational production. In T. Joseph, J. T. Froomkin, D. Jamison, & R. Radner (Eds.), *Education as an industry* (pp. 149-190). Cambridge, MA: National Bureau of Economic Research.

Linn, M.C. (1992). Gender differences in educational achievement. In *Sex equity in educational opportunity, achievement, and testing*. Proceedings of the 1991 Invitational Conference of the Educational Testing Service, Princeton, NJ.

Loeb, S., & Bound, J. (1996). The effect of measured school inputs on academic-achievement: Evidence from the 1920s, 1930s and 1940s birth cohorts. *Review of Economics and Statistics*, *78*(4), 653-664.

Lou, Y., Abrami, P.C., Spence, J.C., Poulesen, C., Chambers, B., & D'Apollonia, S. (1996). Within-class grouping: A meta-analysis. *Review of Educational Research*, *66*(4), 423-458.

Maynard, R., & Crawford, D. (1976). School performance. In D.L. Bawden & W.S. Harrar (Eds.), *Rural income maintenance experiment: Final report* (Vol. 6, Pt. 2, pp. 1-104). Madison: University of Wisconsin, Institute for Research on Poverty.

Michelson, S. (1972). For the plaintiffs—equal school resource allocation. *Journal of Human Resources*, *7*, 283-306.

Monk, D.H. (1992). Educational productivity research: An update and assessment of its role in education finance reform. *Educational Evaluation and Policy Analysis, 14*(4), 307-332.

Murnane, R.J. (1975). *Impact of school resources on the learning of inner city children*. Cambridge, MA: Ballinger.

Odden, A., & Clune, W. (1995). Improving educational productivity and school finance. *Educational Researcher, 24* (9), 6-10, 22.

Pajares, F. (1996). Self-efficacy beliefs in academic settings. *Review of Educational Research*, *66*(4), 543-597.

Perl, L.J. (1973). Family background, secondary school expenditure, and student ability. *Journal of Human Resources*, *8*, 156-180.

Raymond, R. (1968). Determinants of the quality of primary and secondary public education in West Virginia. *Journal of Human Resources*, *3*, 450-470.

*Reynolds, A.J., & Walberg, H.J. (1991). A structural model of science achievement. *Journal of Educational Psychology, 83*, 97-107.

*Reynolds, A.J., & Walberg, H.J. (1992). A structural model of science achievement and attitude: An extension to high school. *Journal of Educational Psychology, 84*, 371-382.

Ribich, T.I., & Murphy, J.L. (1975). The economic returns to increased educational spending. *Journal of Human Resources*, *10*, 56-77.

Robinson, W.S. (1950). Ecological correlations and behavior of individuals. *American Sociological Review, 15*, 351-357.

*Sares, T.A. (1992). *School size effects on educational attainment and ability*. Paper presented at the annual meeting of the American Educational Research Association, San Francisco, CA. (ERIC Document Reproduction Service Number ED 348 743).

Sebold, F. D., & Dato, W. (1981). School funding and student achievement: An empirical analysis. *Public Finance Quarterly*, *9*, 91-105.

Sirotnik, K.A., & Burstein, L. (1985). Measurement and statistical issues in multilevel research on schooling. *Educational Administration Quarterly, 21*(3), 169-185.

Strauss, R.P., & Sawyer, E.A. (1986). Some new evidence on teacher and student competencies. *Economics of Education Review, 5*, 41-48.

Summers, A., & Wolfe, B. (1977). Do schools make a difference? *American Economic Review, 67*, 639-652.

Tuckman, H.P. (1971). High school inputs and their contributions to school performance. *Journal of Human Resources, 6*, 490-509.

Wainer, H. (1993). Does spending money on education help? A reaction to the Heritage Foundation and the *Wall Street Journal*. *Educational Researcher, 22*(9), 22-24.

Walberg, H.J. (1986). Synthesis of research on teaching. In M.C. Wittrock (Ed.), *Handbook of research on teaching* (3rd ed.). Washington, DC: American Educational Research Association.

Walberg, H.J. (1992). The knowledge base for educational productivity. *International Journal of Educational Reform, 1*(1), 5-15.

Wenglinsky, H. (1997). How money matters: The effect of school district spending on academic achievement. *Sociology of Education, 70*, 221-237.

Whiteside, M.F., & Becker, B.J. (1996). *The young child's post-divorce adjustment: A meta-analytic review of the literature.* Report commissioned by the California Judicial Council, State of California.

Winkler, D. (1975). Educational achievement and school peer group composition. *Journal of Human Resources, 10,* 189-204.

[*]Starred references provided data for the example.

Chapter 11

EVALUATION USING SECONDARY DATA

Judy A. Temple

INTRODUCTION

Research using existing data can be a cost-effective means of evaluating educational programs and policies. The availability of many large-scale, longitudinal data sets offers education researchers the opportunity to obtain policy-relevant research findings on the effectiveness of educational policies without going through the expensive and time-consuming process of original data collection. Secondary data analysis can be used to answer new research questions and it can be used to verify the accuracy and robustness of existing research findings.

This chapter describes some important estimation issues that education researchers face when using precollected data to examine the effectiveness of education policies. A number of recently-published education research articles are presented as examples in order to demonstrate the varied nature of the estimation concerns confronting researchers who choose to use data that they themselves had no part in collecting. The broad range of topics considered include the effect of ability grouping on achievement and the effects of compensatory education on educational attainment of children from low-income families. Some recent investigations of the productivity of community college courses, the effect of Catholic

Advances in Educational Productivity, Volume 7, pages 219-239.
Copyright © 1998 by JAI Press Inc.
All rights of reproduction in any form reserved.
ISBN: 0-7623-0253-4

school attendance, and the consequences of frequent school mobility will also be discussed. The examples use data from the National Center on Educational Statistics, the National Science Foundation, as well as other sources.

For the purposes of this chapter, secondary data refers to precollected raw data. Researchers conducting secondary analyses typically rely on data that have been collected by other researchers. In contrast, primary analysis involves estimation using data collected by the researcher explicitly for the study at hand. While it remains typical for education researchers to collect their own data (Reeve & Walberg, 1994), education research using secondary data appears to be growing in popularity. In some sense, education research appears to straddle the middle ground in empirical social science research. While psychologists routinely collect their own data, sociologists and economists have long-standing traditions of analyzing data from surveys or government statistics (Cherlin, 1991; Duncan, 1991).

The researcher's decision to use secondary versus primary data influences the entire research process. In many fields, research is conducted by first formulating precise research questions. Survey or test instruments are selected or designed for use in obtaining the specific measures that the researchers need to test their hypotheses. The experiment is conducted, the data are gathered and then analyzed, and the research report is written. For research relying on secondary data, however, the research questions are generated through a different process. Researchers commonly start with a general area of interest (e.g., ability grouping, school effects on school drop out decisions, the effects of compensatory education) and then investigate the availability of the existing data sets that contain information related to their general interests. The existing data typically have been gathered by other researchers for a different use, or perhaps (as is true with the large-scale national education surveys) they were gathered in order to allow future researchers to ask a wide variety of research questions.

A key distinction between primary and secondary data analysis is that the researchers using the secondary data have to "make do" with the decisions made by others about the precise measures collected and the sampling design used to obtain those measures. Once the secondary researchers have investigated data availability, these researchers commonly revise and narrow the scope of their research questions in order to create precise research questions that can be answered by the data at hand. In the words of McCall and Appelbaum (1991), researchers using primary data follow a linear path in designing and conducting a study. Researchers using precollected data, however, follow a recursive path as research questions are asked and then revised in accordance with the limitations of the existing data. Economists use the phrase "constrained maximization" in explaining how individuals make choices when faced with limited resources. That phrase clearly describes the strategy of secondary data analysts. Researchers using secondary data spend much time trying to figure out the answer to the following question: "What are the best or most appropriate research questions that can be asked given the limitations of the data?"

This chapter adds to a number of existing books, chapters, and articles on secondary data analysis that were written for a broad audience of social science researchers or for a narrower audience of education and/or evaluation researchers. Readers seeking additional information can refer to Boruch (1978), Boruch, Wortman, Cordray and Associates (1981), Hakim (1982), Kiecolt and Nathan (1985), Reeve and Walberg (1994), and Rindskopf (1998). A good overview of research issues facing researchers can be found in a collection of articles published in a 1991 volume of *Developmental Psychology* (including articles by Cherlin, Duncan, and McCall and Appelbaum).

Considered elsewhere in this volume is a type of secondary data analysis research called meta-analysis (Becker, 1998). Meta-analysis is a form of research synthesis in which researchers use the findings of previous studies as data in their own studies. As described by Becker (1998) and by Lipsey (1992), the quantitative integration of research results from earlier studies can help evaluation researchers and policymakers better understand the existing evidence on which educational policies and practices have shown demonstrated effectiveness and which policies and practices have not.

SOME BENEFITS AND COSTS OF SECONDARY DATA ANALYSIS

The main advantage to secondary data analysts of conducting research with pre-collected data is the opportunity to examine a wealth of information on educational processes and outcomes without having to pay the often considerable expenses of primary data collection. Collecting original data requires both money and time. Researchers who rely on secondary data do not have to devote resources to recruiting a sample, constructing survey instruments, administering the survey or experiment, and recording the relevant responses. Education researchers who use secondary data have available to them a vast number of data sets allowing research on topics ranging from education in early childhood to education in graduate school and beyond. Another advantage of secondary data analysis is its accessibility. Because of recent advances in computer technology and software design, research using existing data can be conducted by any trained researcher at any location. The increasing power and decreasing real cost of personal computers and statistical software packages in recent years allows the analysis of secondary data to be undertaken by anyone who has access to a personal computer.

Some additional benefits of secondary data analysis are due to the large number of observations that are included in many large, nationally-representative data sets. The availability of large sample sizes allows researchers to conduct studies of relatively rare events (such as middle-school dropout or teacher burnout) that could not be studied well with smaller samples that contain few observations on the infrequent outcomes. Another benefit of having a large data set is that researchers can investigate the robustness of their results by dividing the sample into two smaller subsamples. As described by McCall and Appelbaum (1991), the statistical analysis could be conducted using a subsample drawn randomly from the larger sample. The researchers would test their hypotheses by developing and applying an appropriate empirical specification to this primary subsample. To assure cross-validity and prevent the researchers from "overfitting" their empirical model, the estimation framework used in the first subsample could then be applied to the remaining sample to see how well the model fits the data. McCall and Appelbaum recommend devoting two-thirds of the original sample for the main analysis and using the remaining one-third of the sample for confirmation purposes. Examples of secondary data analyses in education research that use a second sample for cross-validation purposes include Reynolds and Walberg (1992a, 1992b).

While there exists potential for cost savings through the avoidance of the costs of original data collection, secondary data analysis entails its own, perhaps considerable, costs. An important aspect of secondary data sets is the necessary trade-off of depth for breadth in the availability of measures used to describe education outcomes and processes as well as the characteristics of students, families, teachers, and schools. Researchers using precollected data may be presented with a huge number of variables, but the available measures may be crude approximations of the ideal measures that the researchers would prefer to use. In comparison, researchers who decide to collect their own data in order to test their hypotheses may collect a smaller number of measures. The purposefully selected measures collected by primary data analysts, however, may constitute the best possible information for the purposes of the study at hand.

The direct costs of obtaining secondary data may be low or zero, as the data are often distributed free or at low cost in computer-readable form on tape, diskette, or compact disc. Researchers may have to pay a nominal amount for codebooks. Although these direct costs of obtaining the data may appear low, in practice researchers face other costs that may be considerable. Analysis of both large and small data sets requires taking time to become acquainted with the codebooks and with the physical layout of the individual data items in the large data files. While the large, nationally-representative data sets produced by government agencies and large consulting firms often come with well-written codebooks and other detailed forms of documentation, inadequate documentation is a common complaint of secondary analysts who often have difficulties in deciphering data obtained from other researchers and from smaller agencies and firms. Although many of the large agencies and research firms invest considerable resources in archival activities in order to facilitate future use of the data, individual researchers and small research teams often underinvest in documentation activities. As an example of the seriousness in which data archivists view their work, Robbin (1981) provides detailed guidelines for preparing and documenting data.

In addition to the time costs involved in learning about the codebooks and record layouts that accompany secondary data sets, researchers often have to devote considerable time to learning statistical methods that are appropriate for analysis of the data. Many education researchers receive training in simple analysis of variance (ANOVA) approaches to estimation, for example, but have not received a great deal of training in the types of statistical analysis required for dealing with large data sets that have complex sampling strategies. Duncan (1991) recommends that researchers embarking on a new line of research with an unfamiliar new data set consider teaming up with collaborators who are familiar with the data and the needed statistical tools. In order to learn more about the data and the research design issues associated with a particular secondary data set, researchers also are encouraged to read papers by others who have used the same data even if the topics addressed by the previous researchers are not related to the current researcher's interests.

Secondary analysis entails considerable costs of learning about the intricacies of the data and of acquiring the statistical tools required for analyzing large-scale national and international data sets. In recognition of these costs, the American Educational Research Association (AERA) has in recent years sponsored training for researchers at an Institute for Statistical Analysis for Education Policy. The Institute also has been supported by the National Center for Education Statistics (NCES) and the National Science Foundation (NSF). Additional AERA funding in recent years has been devoted to a small grants program that supports researchers who use the nationally-representative education data sets produced by either NCES or NSF.

One important concern to education policymakers is the perceived underutilization of the data set called the National Assessment of Educational Progress (NAEP), which has been described by the National Research Council as "an unparalleled source of information about the academic proficiency of U.S. students..." (U.S. Department of Education, 1997). NAEP biannually assesses student performance at the fourth, eighth and twelfth grades in a number of subjects, and the NAEP data offer invaluable information on long-term trends as well as cross-sectional national and state-by-state comparisons of student populations of interest. Due in part to the complexity and the quantity of the NAEP data, the data are infrequently used by researchers outside of the federal government. In an effort to stimulate research using NAEP, the National Center for Education Statistics offers special data extraction programs and an SPSS module that help researchers access the relevant data and carry out the complicated analyses necessary due to the complex sampling and assessment mechanisms used in the NAEP design (U.S. Department of Education, 1997).

Because of the tremendous investment of time required to conducting work with a particular complex data set, it is no surprise that we commonly observe education researchers using the same secondary data set as the basis for a number of research papers instead of using a different data set for each research project.

Because of the complexities associated with any one large data set, it is likely that researchers who take the time to specialize in and master the details of a relatively small number of secondary data sets may produce better quality research than other researchers who may choose to use a larger number of secondary data sets without becoming an expert in any one of them.

RESEARCHERS' EXPERIENCES WITH REPLICATION

To verify findings of other researchers, secondary data analysts occasionally use existing data in an attempt to replicate the findings of earlier studies as closely as possible. Although relatively infrequent in social science research, replication activities are an important part of the scientific process (Campbell, 1994). Successful replication of previous findings lends strong support to the validity of those findings. Replication attempts can also point out that the original findings are incorrect due to mistakes in data entry, variable transformations, and statistical analysis.

The results of an investigation of replication in economics may be of some interest to other researchers considering replication activities. In the early 1980s, researchers funded by the National Science Foundation embarked on a two-year study that attempted to replicate published results in a highly-regarded economics journal that published empirical papers in the area of banking and monetary economics (Dewald, Thursby, & Anderson, 1986; Anderson & Dewald, 1994). Contributors to the journal were required to submit their programs and data to the archives maintained by the editors. The replication researchers wanted to ascertain whether or not the findings of the published studies could be replicated from the authors' own programs and data sets, and also whether it was possible to replicate findings by going directly to the data sources mentioned in the articles. Disappointedly common were instances in which the data were not sufficiently well-labeled and the provided programs not sufficiently detailed to allow for successful replication. In the cases in which the original researchers provided sufficient program and data information to allow replication to take place, the results of the replication experiments suggested that inadvertent errors in empirical research (such as mislabeled or incorrectly created variables and data entry mistakes) are common. In many instances, however, the errors did not affect the main conclusions of the studies.

Secondary data analysts sometimes conduct reanalysis of existing program and policy evaluations in order to verify the accuracy of earlier findings and to see how robust the earlier findings are to changes in the estimation methods used in the analysis. An early and important example of an educational evaluation that stimulated much subsequent reanalysis was the controversial Westinghouse evaluation of the Head Start program (Cicirelli et al., 1969). Subsequent research by many social science researchers (including Barnow & Cain, 1977;

Magidson, 1977; and Bentler & Woodward, 1978) has helped clarify the inferences of the earlier studies and has helped researchers become more aware of the statistical methods appropriate for evaluations based on quasi or nonexperimental designs. In particular, researchers have learned important lessons from the early Head Start studies on the importance of being able to control for differences in the characteristics of students in treatment and control groups. The original Head Start evaluation data continue to be analyzed almost 30 years later by researchers using newer estimation methods and examining different subgroups of the original sample (e.g., Wu & Campbell, 1996).

Robert Boruch and his colleagues (Boruch et al., 1981) have devoted much attention to reanalysis of evaluations of a variety of social and educational programs and policies. Their reanalyses confirmed results of some earlier studies, cast doubt on the findings of some other evaluations, and suggested improvements in research methods. Just as in the case of the economics replication experiments, the researchers involved in Boruch and Wortman's Project on Secondary Analysis (Boruch & Wortman, 1978) on some occasions encountered difficulties in replicating the findings of previous researchers. For example, Linsenmeier, Wortman, and Hendricks (1981) were unable to replicate published findings on the relationship between teachers' racial biases and student performance using data from the Riverside School Study of Desegregation (Gerard & Miller, 1975). Linsenmeier and colleagues found the Riverside data files to be very well documented. The problem, however, was the lack of documentation for the computer programming used to transform the data into useful measures and to obtain the relevant sample size. Even after consultation with some of the original researchers, the secondary analysts were unable to determine how the original sample size was determined or how many key variables were calculated.

It is possible that recent developments in statistical software for personal computers may in the future result in more complaints by would-be replicators of inadequate program documentation. Instead of requiring analysts to submit clearly written syntax in complete programs, recent Windows-based versions of widely used statistical software packages including SPSS and SAS now allow new variable creation and extensive statistical analyses to be performed with the click of a mouse button instead of through submission of a explicitly-written program. While the precise steps undertaken in these point-and-click estimation sessions can be retrieved in the form of a log at the end of each estimation session, it is possible that the original researchers may not retain the documentation. As a consequence, the researchers may not retain sufficiently detailed evidence of the steps through which the estimation was conducted to allow themselves or others to retrace their steps. Both primary and secondary data analysts may want to consider continuing to use the computer's syntax language to write explicit programs for analyzing the data, or at least make a regular practice of retaining in an organized way the log record of each estimation session.

Although replication activities are not a required component of secondary data analysis, replication can be useful for two main reasons. First, replication is crucial for verifying the accuracy of research results. Verification by other researchers is especially crucial when evaluation results are controversial and the results may be used as the grounds for making public policy. A recent example of replication undertaken to verify controversial findings is the work of Kane and Rouse (1995) in determining whether or not individuals who enroll in community college courses end up earning higher subsequent wages than otherwise similar individuals who did not enroll in courses beyond high school. Kane and Rouse were attempting to verify the accuracy of the findings of Grubb (1993), whose research found no benefit in terms of higher wages for students who had attended community colleges. After attempting to replicate Grubb's findings, Kane and Rouse (1995) determined that Grubb had made programming mistakes that affected the conclusions that could be drawn from his study. After fixing these mistakes, Kane and Rouse concluded that Grubb's conclusions in his 1993 study were unwarranted, as enrollment in community college classes either with or without eventually earning a degree was associated with higher subsequent wages.

Another benefit of attempting at least a partial replication of earlier studies is that replication is a good way for a researcher to become familiar with a particular data set. Hands-on experience in figuring out what decisions other researchers have made with respect to variable transformations and sample size considerations can be especially useful when the secondary analysts are attempting to extend and improve upon previous work using the same data.

METHODOLOGICAL ISSUES

Although there are not likely to be any estimation issues that arise in the analysis of secondary data that could not arise with primary data, researchers using precollected data are especially likely to be faced with some important methodological concerns. Before undertaking any statistical analysis of a data set, secondary analysts must acquaint themselves with some basic features of the original study design. Some design features discussed here include the sampling design, the availability of measures that the secondary analyst may view as less than ideal, and sample selection biases that may result due to nonrandom attrition or the nonrandom assignment of educational policies or programs across different groups. This discussion of the methodological issues is not meant to be exhaustive. Rather, some important issues are discussed in some detail, while many other issues such as missing data and measurement error are left unaddressed. For more information, the reader can consult Kiecolt and Nathan (1985), Little and Rubin (1987), and Reeve and Walberg (1994).

Sampling

Knowledge of the sampling design used by the original collectors of the data is important for proper use of the secondary data set and proper interpretation of the study findings. Large data sets intended to be representative of an even larger population typically include *sampling weights*. For the sample to be truly representative of the target population, members of small demographic groups are oversampled. Without intentionally oversampling, random draws from a large population would result in a very small number of observations on members of certain minority groups, and the concern is that the few minority group members actually included in the sample might not be representative of the population.

Sampling weights, which adjust for unequal probabilities of selection into the sample, are used to reduce the importance of the oversampled observations in the sample statistics so that the weighted sample is both representative of the population and includes sufficient information on minority group members. Secondary analysts use the sample weights in calculating descriptive statistics and in conducting additional statistical analyses. In some research projects, however, analysts ignore the sample weights and conduct their research using the unweighted, unrepresentative data. One reason may be that some estimation procedures do not easily allow incorporation of the weights. Rumberger (1995) uses a hierarchical linear modeling (HLM) approach to looking at the determinants of middle school dropout and cautions that his student-level estimates in HLM are not based on the weighted sample because his estimation procedure is unable to take the weights into account. In criticizing the conclusions drawn by previous researchers in evaluations of job-training programs, Heckman, Hotz, and Dabos (1987) argue that some simple estimators used by the previous researchers are not appropriate for samples in which population members have unequal probabilities of selection into the sample unless the weights are explicitly incorporated into the estimation methods. Occasionally, education researchers report without explanation that they have chosen not to use the sample weights (e.g., Evans & Schwab, 1995). Estimation using the unweighted observations yields results that may not be generalizable to the population as a whole.

In addition to taking into account the sample weights, secondary analysts need to pay attention to another aspect of the sampling design that arises when there is *multi-level sampling*. For example, the National Assessment of Educational Progress (NAEP) is based on an especially elaborate sampling design. In order to obtain a nationally representative sample, the sampling procedure for NAEP begins with the selection of primary sampling units, which are geographic areas consisting of single or multiple counties. These PSUs are selected by stratifying the units according to region of the country, urbanicity, student ethnicity, educational attainment and the median income of the local residents. Once the PSUs are chosen, schools are selected next, followed by the selection of students within schools. Although other large-scale education surveys may use a less complicated sampling design, it is common that schools are selected in some way from the entire population of schools and then students are randomly selected within schools. Estimation strategies that take the sequential sampling into account are important because the students within each school cannot be considered to be independent observations. As explained by Raudenbush (1988, p. 85), "The traditional linear models on which most researchers rely require the assumption that subjects respond independently to educational programs. In fact, subjects are commonly 'nested' within classrooms, schools, districts, or program sites so that the responses within groups are dependent."

Researchers from a number of disciplines have recognized the importance of taking into account the nonzero intra-group correlation among individuals that arises in nested or grouped data structures. The relevant estimation approaches are called "hierarchical" modeling by education researchers, "random coefficient" modeling by economists, "multi-level" modeling by sociologists, and "mixed" modeling by researchers in a variety of disciplines. A popular statistical package used by education researchers is called Hierarchical Linear Modeling (Bryk & Raudenbush, 1992). Medical researchers and some social scientists commonly use MIXREG or MIXOR (Hedeker & Gibbons, 1996). Recently, it has become possible to use SAS, a widely-available multi-purpose statistical package, to conduct estimation with nested data structures using a procedure called SAS PROC MIXED (Wang, 1997).

Hierarchical data structures can arise quite frequently, even when the sampling is not explicitly randomized at multiple stages. For example, neighborhood and school characteristics might be merged into a data set containing information on individual student achievement. Studies that try to explain variation across individual micro-units (e.g., individual students) in terms of macro-level variables (e.g., neighborhood and school characteristics) are commonly referred to as *contextual study designs*. As discussed by Kiecolt and Nathan (1985), in a contextual design the behavior of an individual residing in or attending a neighborhood or school is explained by a combination of micro- and macro-level variables. The micro variables refer to individual-specific characteristics and the macro variables represent characteristics of the neighborhood or school. In a contextual design, many individuals may share the same macro-level characteristics. The standard estimation procedures such as ordinary least squares assume that individual observations are independently drawn from a larger population. In the case of these hierarchical study designs, the individuals within each group cannot be considered to be independent. For example, a sample of 1000 students nested into 20 schools cannot be considered to be 1000 independent observations because the students within each school are likely to share a common variance. Ordinary least squares regression of school-level macro variables on individual-level measures of achievement often underestimates the standard errors associated with the macro-level variables. OLS estimation assumes that there is more information on the

1000 individuals than really exists because in reality there are not 1000 independent observations.

An eye-catching non-education example that well illustrates the general problem with merging macro-level data with observations on micro-level units is provided by Moulton (1990). In an attempt to explain the variation in wages across individual workers who reside in many states across the country, Moulton includes as predictors of individual wages some state-level characteristics that are common for all workers within a state but vary across states. To illustrate that the uncorrected t-statistics associated with the macro-level variables are likely to be biased upward, he also includes as a predictor of individual wages a random number that is the same for all workers within a state but varies across states. Without the correction of the standard errors required as a result of the nested nature of the data, Moulton finds that a number of state-level variables, including the random number, are highly significant predictors of individual wages. With the proper correction that takes into account the nonzero intra-state correlation in the individual errors, the estimated coefficients on many of the state-level variables (including the random number) are insignificantly different from zero.

Existence of Appropriate Measures

A potentially serious problem with the use of precollected data is that the available measures may be crude approximations of the precisely-defined measures the analyst would prefer. The absence of measures that will satisfy every user of the large national surveys intended for public use is not surprising, or course, given that these surveys are designed to allow many users to be able to ask a large number of research questions. The absence of the "ideal" measures may affect the definition of the dependent or the explanatory variables, and the researcher should be upfront about the limitations of the data. Some obvious examples include the availability of appropriate measures of teacher quality and of the student's socioeconomic status, as both teacher quality and family income are often hypothesized to affect student achievement. Teacher quality often is represented by the number of teachers with M.A. degrees and with the average number of years of teacher experience. But high numbers of experienced teachers and graduate degree-holding teachers might be better proxies for teacher age than for teacher quality. A common measure of student socioeconomic status is a dummy variable reflecting eligibility for a subsidized school lunch. While the subsidy eligibility is a function of income, use of this measure instead of actual family income certainly prevents the researcher from taking into account the great income differences that may exist among students within the lunch eligibility categories.

The timing of the measurements in secondary data sets may not be ideal either. In some studies, individuals are asked to recall aspects of their education or other events that took place years ago. For example, in a study of the effect of the compensatory preschool program Head Start on child health, Currie and Thomas

(1995) use information from the National Longitudinal Survey of Youth (NLSY) in which mothers were asked to recall if their children had ever received measles immunizations. Wood and associates (1993) investigate the effect of frequent school mobility by using information from the National Health Insurance Survey in which students were asked to recall how many schools they previously had attended. Obviously, the use of retrospective rather than prospective study designs may yield inaccurate information on important variables.

Another aspect of the timing issue that arises even in prospective study designs is the existence of a less-than-ideal time interval over which to observe the variables of interest. One example is a study of the predictors of frequent school mobility conducted by Temple and Reynolds (1997) using data from the Chicago Longitudinal Study. Although Temple and Reynolds use measures of school mobility that are obtained through school records and consequently do not have to rely on problems with faulty recollections, their measure of mobility is affected by the time frame over which data are gathered. Temple and Reynolds measure school mobility by examining the school codes of each student obtained annually from administrative records. Students whose school codes changed from one year to the next were assumed to have moved exactly once during that year. As explained by the authors, the measure of the "number of school moves" is more appropriately considered a measure of the "number of years in which the student changed schools at least once." The "window problem" of the non-ideal timing of the observations on student-level information used to estimate the determinants of educational attainment recently has been examined in detail by Wolfe, Haveman, and An (1996).

Selection Biases Due to Nonrandom Program Assignment and to Attrition

In evaluating the effectiveness of educational policies or programs, researchers rarely, if ever, are able to observe a random sample of a population of interest. Even if units of observation such as students, teachers, or schools are randomly selected from the larger population, it is likely that some of their important characteristics cannot be viewed as randomly determined. In education research about the relative productivity of private versus public schools, for example, students are not randomly assigned to public and private schools. In studies about the effect of ability grouping on student achievement, students are not likely to be randomly assigned into tracks. In studies of the effects of frequent school mobility on subsequent student achievement, frequent school mobility is never randomly assigned across students. A key problem in making inferences about the "effect" of educational or other variables on educational outcomes is that the explanatory variables (e.g., private school placement, tracking, or frequent mobility) may not be independent of the error term in an equation used to explain the variation across students in the educational outcomes. In other words, there may be unobserved student or family characteristics that are correlated with both the educational outcome and the decision to attend private school, as an example. If the students who attend private schools have unmeasured characteristics (such as motivation or especially supportive parents) that would have lead them to do well in either private or public schools, then estimating the effect of private school attendance without taking into account this positive selection or "creaming" problem would yield an estimate of the effect of private school attendance that would be biased upward.

Nonrandom attrition is a related concern that faces researchers who use longitudinal data. Secondary data sets using individuals as units of observation almost always suffer from some data loss as the original researchers lose track of some individuals over time. Nonrandom attrition can result in biased inferences if each individual's probability of remaining in the sample is a function of the error term in the outcome equation. For example, it may be the case in studies analyzing the determinants of achievement gains from one year to the next that some students with test scores in year one have left the sample before taking the test in year two. In some cases, failing to control for nonrandom attrition can lead to incorrect inferences about the effects of educational policies on student outcomes. The important issue is whether attrition is random or at least random conditional on the included explanatory variables, or whether there are unobserved student characteristics that are correlated with both the educational outcome and the probability that a particular student leaves the sample.

As described by Rindkopf (1998, 1992), both types of selection problems described above can be considered missing data problems. For attrition, of course, data on the dependent variable obviously are missing. Viewing as a missing data problem the nonrandom assignment of important explanatory variables such as private school attendance, ability grouping, or frequent mobility requires a little more explanation. In order to estimate the effect of private school attendance on achievement, for example, the researcher ideally would like to know what the student's achievement would be under two states of the world. One state is the observed state—the level of achievement attained by a private school student. The other state is often referred to as the *counterfactual*. The counterfactual in this case is the achievement of the private school student if that student had attended public school instead. The effect of private school attendance on achievement for that student is the simple difference in the achievement level attained in private school minus the achievement level that would have occurred if the student had attended public school. Obviously, what is missing here is information on the counterfactual—we do not know the achievement level that would have occurred if the private school student had attended public school instead.

The most popular approaches to explicitly addressing both types of selection concerns come from the econometrics literature and typically rely on Heckman (1979) sample selection methods or other related techniques to control for both nonrandom program assignment or nonrandom attrition. Both attrition and program assignment concerns are addressed by explicitly modeling either the probability of attrition or the student's decision to enroll in private school (or, continuing the examples from above, to be placed in tracked classes or to experience frequent mobility) along with an equation for the outcome of interest.

A good discussion of the use of Heckman's sample selection method for controlling for attrition is provided by Becker and Walstad (1990). Becker and Walstad estimate the effect of a tutoring program on economics knowledge as measured by a standardized test called the TUCE. The researchers employ the commonly used value-added test score equation in which the dependent variable is the student's TUCE test score. The explanatory variables include a dummy variable equal to one if the student had participated in the tutoring classes and the student's TUCE score at an earlier point in time before the tutoring sessions were offered. Because a non-trivial number of students in their sample took the TUCE the first time but not the second, the students with a missing second test score cannot be included in the value-added equation. Becker and Walstad estimate an attrition equation predicting which students are thrown out of the sample. Information from that equation is used to construct a new variable called the inverse mills ratio (or Heckman's lambda), and that new variable is added as an explanatory variable in the outcome equation in an attempt to control for the unobservable differences between students who left the sample and students who remained. Becker and Walstad demonstrate that controlling for nonrandom attrition is important because the estimated effect of the tutoring program is biased when the attrition problem is ignored.

More recent research including Reynolds and Temple (1998) employs a newer form of the two-step Heckman selection correction method that controls for nonrandom attrition by jointly estimating an equation for attrition with an outcome equation. Instead of constructing the inverse Mills ratio from the first equation and then inserting it as an additional covariate in the outcome equation, the newer maximum likelihood version of this selection correction method controls for unobservable student characteristics that may predict both attrition and the outcome by allowing the error terms in the two equations to be correlated.

Two-equation methods for addressing selection concerns also are commonly used when there is reason to believe that the assignment of educational policies or programs is not random across students. Recent studies estimating the effect of Catholic school attendance, for example, commonly estimate a separate equation predicting which students enroll in private versus public schools (e.g., Neal, 1997; Evans & Schwab, 1995; Murnane, Newstead, & Olsen, 1985). Popular two-equation estimation methods include the Heckman sample selection method and instrumental variables procedures. As explained by Murnane and associates, the use of two-equation methods for controlling for selection biases due to either nonrandom attrition or nonrandom program assignment generally requires that researchers find at least one variable (often called an "instrument") that predicts attrition or program assignment but does not predict the academic outcome. In the Catholic versus public school studies, researchers have employed a variety of instruments including proximity to Catholic schools and religious affiliation of students. As described in Neal (1997), a number of recent studies find evidence of *negative* selection into Catholic schools, suggesting that students with unobserved traits that would lead them to be less successful students are actually more likely to enroll in Catholic schools. Consistent with the earlier findings by Coleman, Hoffer, and Kilgore (1982), the recent studies that control for sample selection on unobserved student characteristics find that Catholic school attendance has a positive effect on educational attainment.

Readers interested in learning more about estimation methods useful for controlling for selection biases due to nonrandom program assignment or nonrandom attrition and the assumptions that underlie each of the methods can consult Winship and Mare (1991), Cook and Shadish (1994), Tuijnman (1994), Reynolds and Temple (1995), Foster and Bickman (1996), and Rindskopf (1998 and 1992).

A Brief Listing of Some Available Secondary Data Sets

Although secondary data can be obtained from a variety of sources, two U.S. government agencies specialize in collecting and disseminating data on a wide range of education topics. The National Center for Education Statistics (NCES), which is located in the U.S. Department of Education, conducts a large number of surveys and makes data available to researchers. These surveys include elementary and secondary surveys such as the Common Core of Data (CCD) and the Schools and Staffing Survey (SASS). The CCD contains annual information on demographics, revenues, and expenditures for all public elementary and secondary schools in the United States. The NCES obtains these data from state education departments. The SASS data on school and teacher characteristics are drawn from a sample of public and private schools, and are or will be available for the 1987-1988, 1990-1991, 1993-1994, and the 1998-1999 school years. The NCES also conducts post-secondary surveys, longitudinal surveys, national education assessments, and surveys of libraries. The National Science Foundation (NSF) conducts a number of surveys that focus on the mathematics and science education of students and teachers, the training and work histories of engineers and scientists, and research and development expenditures.

Due to the large number of existing data sets, no attempt is made here to summarize or even briefly describe the type of data contained in each data set. Table 11.1 contains a partial list of additional NCES or NSF data sets that are available to researchers. More information on these data sets can be obtained through the NCES or NSF offices and through the relevant internet sites.

Table 11.1. A Partial List of Secondary Data
Sets for Education Evaluation Researchers

Data set	Conducted or funded by[*]
National Household Education Survey (NHES)	NCES
National Education Longitudinal Study of 1988 (NELS:88)	NCES
High School and Beyond (HS&B)	NCES
Early Childhood Longitudinal Study: Kindergarten Class of 1998-99 (ECLS)	NCES
National Assessment of Education Progress (NAEP)	NCES
Third International Mathematics and Science Study (TIMSS)	NCES and NSF
Longitudinal Study of American Youth (LSAY)	NSF
National Survey of Recent College Graduates (NSRGG)	NSF
Survey of Graduate Students and Postdoctorates in Science and Engineering (GSS)	NSF

Note: * readers can obtain more information from the following internet addresses:
www.nces.ed.gov; www.nsf.gov; www.lsay.org.

EVALUATION OF EDUCATIONAL POLICIES
USING SECONDARY DATA: A RECENT EXAMPLE

Throughout this chapter, examples have been presented of education research that uses secondary data to evaluate the effectiveness of educational policies and programs. Areas of research very briefly touched upon included the effects on teacher biases on student performance, the effects on wages of taking courses at community colleges, and the effects on achievement of compensatory preschool education programs, Catholic school attendance, or frequent school mobility. Here, more information is provided about recent research in one particular topic area: the effect of ability grouping on student achievement. The choice of the specific articles is guided by the desire to provide fairly recent examples of how education researchers have evaluated educational policies using different data sources. Readers seeking more information about research on ability grouping are encouraged to read the papers cited and consult the list of references therein, as well as the discussion of ability grouping in the chapter in this volume by Boruch and Terhanian (1998).

Example—Evaluating the Effectiveness of Ability Grouping on Achievement

Researchers have recently used large secondary data sets to examine the prevalence and effects of the school practice of ability grouping. Hoffer (1992) uses data from the Longitudinal Study of American Youth (LSAY) to investigate the effect of ability grouping in mathematics and science on achievement growth from seventh to ninth grade in a sample of middle school students. More recently, Rees, Argys, and Brewer (1996) and Argys, Rees, and Brewer (1996) have used data from the National Longitudinal Study of 1988 (NELS) to examine the prevalence of ability grouping in a number of subjects and the effect of ability grouping in mathematics courses on math achievement.

Both Hoffer (1992) and Rees and associates (1996) recognize the imperfect nature of the available measures of "tracked" classes. Both the LSAY data (funded by the National Science Foundation) and the NELS data (funded by the National Center on Education Statistics) were explicitly constructed to allow researchers to investigate tracking decisions and consequences along with many other education research questions. Teachers in the LSAY study were asked, "Are the students in your school grouped by ability or prior achievement either as a result of student choice or school policy in the 7th grade (8th grade) science (mathematics) program?" Recognizing that not all teachers may agree on the definition of tracking or ability grouping, the researchers responsible for creating the LSAY data set followed up these teacher surveys in some cases with phone calls to the school and also gathered curriculum information from each school to get a clear picture of the prevalence of tracking in math and science middle school classes. The LSAY findings indicate that ability grouping in middle school is common in mathematics but not in science.

As explained by Rees and associates (1996) the NELS data contain both a formal and informal measure of ability grouping. The formal measure is obtained from information about whether the tenth grade courses taken by each sampled student are considered by the students' teachers to be honors, academic, general, vocational, or "other" track. The informal measure of tracking in eighth and tenth grade was obtained by the responses given by teachers when asked how the students in their classes compared to average student in the school. Specifically, teachers were asked to state if the students in each of their classes had higher, average, or lower achievement levels than the typical student in the school, or whether their classes contained students of "widely differing achievement levels." The informal definition of tracking considers an untracked class to be a class in which there are students of widely differing achievement levels. Both the formal and informal measures indicate that tracking is prevalent in math, science, English, and social studies courses.

Both Hoffer (1992) and Argys and associates (1996) investigate the effect of being placed in high, middle, low, or untracked courses on achievement growth using a value-added test score equation in which individual student achievement is regressed on a number of variables including earlier achievement, dummy variables representing track assignment, and other variables designed to control for observed differences in the characteristics of students assigned to tracked and untracked classes. Both studies pay much attention to the possibility of selection bias tainting the estimated "effects" of tracking since tracking cannot be viewed as being randomly assigned across students. Researchers in both studies employ two-equation methods to control for the possibility that there might be unobserved student characteristics that may be correlated with both the tracking decision and student achievement.

The studies by Hoffer (1992) and Argys and associates (1996) use different data sets, different measures of tracking, and different sets of control variables. In particular, Argys and associates control for a wider range of teacher and school characteristics that may influence track assignment and student achievement. Both studies employ Heckman-type sample selection corrections involving the estimation of two equations—one for the achievement outcome and one for track placement. Both studies along with Rees and associates (1996) recognize the non-ideal definition of "ability grouping" found in their chosen data sets and provide sufficient information about the definition of the tracking measures found in the LSAY and NELS data to allow readers to come to their own conclusions about the appropriateness of the measures. Both studies make a point of addressing potential biases due to nonrandom attrition and conclude that controlling for attrition does not affect the estimates. Both studies employ the sampling weights that are designed to ensure the representativeness of the sample statistics. However, as is common in studies that focus on sample selection concerns, neither study explicitly addresses the nested study designs in which students are randomly selected within schools in both the LSAY and the NELS data.

Despite the different data sets, both Hoffer and Argys and associates come to the same conclusion, and their findings are especially important to education researchers because they contradict the often-cited findings of Slavin (1990). Based on his review of the literature, Slavin concluded that the effect of tracking was negligible for students of all achievement levels. The recent research of Hoffer (1992) and of Argys, Rees, and Brewer (1996) demonstrate that ability grouping benefits students placed in the high-ability tracks and harms students placed in the low-achievement tracks. The findings suggest that a nationwide policy of eliminating tracking would benefit students who are low-achievers but harm students who are high-achievers.

SUMMARY AND CONCLUSION

Many existing data sets contain a wealth of information about education policies. These secondary data sets are available to researchers to conduct policy-relevant evaluations of the effectiveness of educational policies and programs without going through the trouble of original data collection. While secondary data analysis can be used to answer new research questions, an important strength of secondary analysis is its tremendous usefulness in verifying the accuracy and robustness of previous research findings. Researchers using secondary data can provide education policymakers with important information about the validity of previous policy findings either through explicit attempts at replicating previous work, often with new statistical techniques, or by testing similar hypotheses using different data sets in order to see how robust earlier research findings are across different data sets and estimation methods. The net result of these analyses is to increase confidence in research findings and their implications for programs and policies.

This chapter has identified some common estimation concerns that confront researchers who use data that they themselves had no say in collecting. One important concern is the need for the users of the data to assess the suitability of the existing measures of the educational policies and outcomes of interest. Secondary data sets often are rich in terms of the large number of variables collected. To researchers with very specific needs for precisely-defined measures of educational policies or outcomes, however, the available measures may be crude approximations of the preferred measures. Another estimation concern facing researchers who use large secondary data sets is the need for estimation skills suitable for handling complex sampling designs as well as the causal inference problems that arise because of the nonrandom assignment of educational policies across students or schools. As a guide for readers who would like to read more about the use of secondary data analysis in education evaluation research, this chapter identifies a number of recent examples of education research conducted through reliance on secondary data. Stakeholders of educational programs and policies can benefit from the greater use by researchers of existing data sets.

REFERENCES

Anderson, R. G. & Dewald, W. G. (1994). Replication and scientific standards in applied economics a decade after the Journal of Money, Credit & Banking project. *Federal Reserve Bank of St. Louis Review, 76*, 79-83.

Argys, L. M., Rees, D. I., & Brewer, D. J. (1996). Detracking America's schools: Equity at zero cost? *Journal of Policy Analysis and Management, 15*(4), 623-645.

Barnow, B. S., & Cain, G. C. (1977). A reanalysis of the effect of Head Start on cognitive development: Methodology and Empirical Findings. *Journal of Human Resources, 12*, 177-197.

Becker, B. (1998). Research synthesis in the study of educational productivity. In A. Reynolds & H. Walberg (Eds.), *Evaluation research for educational productivity*. Greenwich, CT: JAI Press.

Becker, W. E. & Walstad, W. B. (1990). Data loss from pretest to posttest as a sample selection problem. *Review of Economics and Statistics, 72*, 184-188.

Bentler, P. & Woodward, J.A. (1978). A Head Start reevaluation: Positive effects are not yet demonstrable. *Evaluation Quarterly, 2*, 493-510.

Boruch, R. F., & Terhanian, G. (1998). Cross design synthesis, In A. Reynolds & H. Walberg (Eds.), *Evaluation research for educational productivity*. Greenwich, CT: JAI Press.

Boruch, R. F., Wortman, P. M. Corday, D. S. & Associates (1981). *Reanalyzing program evaluations: Policies and practices for secondary analysis of social and educational programs*. San Francisco, CA: Jossey-Bass, Inc.

Boruch, R. F. (Ed.). (1978). *New directions for program evaluation: Secondary analysis*. San Francisco, CA: Jossey-Bass, Inc.

Boruch, R. F. & Wortman, P. M. (1978). An illustrative project on secondary analysis. In R.F. Boruch (Ed.), *New directions for program evaluation: Secondary analysis* (pp. 89-110). San Francisco, CA: Jossey-Bass, Inc.

Bryk, A., & Raudenbush, S. (1992). *Hierarchical linear models: Applications and data analysis methods*. Newbury Park, CA: Sage Publications.

Campbell, D. T. (1994). Retrospective and prospective on program impact assessment. *Evaluation Practice, 15*, 291-298.

Cherlin, A. (1991). On analyzing other people's data. *Developmental Psychology, 27*(6), 946-948.

Cicirelli, V.G., & Associates (1969). *The impact of Head Start: An evaluation of the effects of Head Start on children's cognitive and affective development, Vol. I and II*. Athens: Westinghouse Learning Corporation and Ohio University.

Coleman, J., Hoffer, T. & Kilgore, S. (1982). *High school achievement: Public, Catholic, and private schools compared*. New York: Basic Books.

Cook, T. D., & Shadish, W. R. (1994). Social experiments: Some developments over the past fifteen years. *Annual Review of Psychology, 45*, 545-580.

Currie, J., & Thomas, D. (1995). Does Head Start make a difference? *American Economic Review*, 341-364.

Dewald, W. G., Thursby, J. G. & Anderson, R. G. (1986). Replication in empirical economics: The journal of money, credit & banking project. *American Economic Review, 76*, 587-603.

Duncan, G. J. (1991). Made in heaven: Secondary data analysis and interdisciplinary collaborators. *Developmental Psychology, 27*(6), 949-951.

Evans, W., & Schwab, R. (1995). Finishing high school and starting college: Do Catholic schools make a difference? *Quarterly Journal of Economics, 110*, 947-974.

Foster, E. M. & Bickman, L. (1996). An evaluator's guide to detecting attrition problems. *Evaluation Review, 20*, 695-723.

Gerard, H.B., & Miller, N., (Eds.). (1975). *School desegregation*. New York: Plenum Press.

Grubb, W. N. (1993). The varied economic returns to postsecondary education: New evidence from the class of 1972. *Journal of Human Resources, 28*, 365-382.

Hakim, C. (1982). *Secondary analysis in social research: A guide to data sources and methods with examples*. London: Allen & Unwin.

Heckman, J. J., Hotz, V. J., & Dabos, M. (1987). Do we need experimental data to evaluate the impact of manpower training on earnings? *Evaluation Review, 11*, 395-427.

Heckman, J. J. (1979). Sample selection as a specification error, *Econometrica, 47*, 153-161.

Hedeker, D., & Gibbons, R. D. (1996). MIXREG: A computer program for mixed-effects regression analysis with autocorrelated errors. *Computer Methods and Programs in Biomedicine, 49*, 229-252.

Hoffer, T. B. (1992). Middle school ability grouping & student achievement in science and mathematics. *Educational Evaluation and Policy Analysis, 14*, 205-227.

Kane, T. J., & Rouse, C. E. (1995). Comment on W. Norton Grubb. *Journal of Human Resources, 30*, 205-221.

Kiecolt, K. J., & Nathan, L. E. (1985). *Secondary analysis of survey data*. Beverly Hills, CA: Sage Publications, Inc.

Linsenmeier, J. A. W., Wortman, P. M. & Hendricks, M. (1981). Need for better documentation: Problems in a reanalysis of teacher bias. In R. F. Boruch, et al. (Eds.), *Reanalyzing program evaluations: Policies and practices for secondary analysis of social and educational programs* (pp. 68-83). San Francisco, CA: Jossey-Bass, Inc.

Lipsey, M. W. (1992). Meta-analysis in Evaluation research: Moving from description to explanation. In H. Chen & P. H. Rossi (Eds.), *Using theory to improve program and policy evaluations* (pp. 229-241).

Little, R., & Rubin, D. B. (1987). *Statistical analysis with missing data*. New York: Wiley.

Magidson, J. (1977). Toward a causal model approach for adjusting for preexisting differences in the nonequivalent control group situation: A general alternative to ANCOVA. *Evaluation Quarterly, 2*, 511-520.

McCall, R. B., & Appelbaum, M. I. (1991). Some issues of conducting secondary analyses. *Developmental Psychology, 27* (6), 911-917.

Moulton, B. R. (1990). An illustration of a pitfall in estimating the effects of aggregate variables on micro units. *Review of Economics and Statistics, 72*, 334-338.

Murnane, R. J., Newstead, S., & Olsen, R. J. (1985). Comparing public and private schools: The puzzling role of selectivity bias. *Journal of Business and Economic Statistics, 3*, 23-35.

Neal, D. (1997). The effects of Catholic secondary schooling on educational achievement. *Journal of Labor Economics, 15*, 98-123.

Raudenbush, S.W. (1988). Educational applications of hierarchical linear models: A review. *Journal of Educational Statistics, 13*, 85-116.

Rees, D. I., Argys, L. M., & Brewer, D. J. (1996). Tracking in the United States: Descriptive statistics from NELS. *Economics of Education Review, 15*(1), 83-89.

Reeve, R.A., & Walberg, H. J. (1994). Secondary data analysis. In T. Husen & T. N. Postlewaite (Eds.), *The international encyclopedia of education* (2nd ed., Vol. 9).

Reynolds, A. J., & Temple, J. A. (1998). Extended early childhood intervention and school achievement: Age 13 findings from the Chicago longitudinal study. *Child Development, 69*, 231-246.

Reynolds, A. J., & Temple, J. A. (1995). Quasi-experimental estimates of the effects of a preschool intervention. *Evaluation Review, 19*, 347-373.

Reynolds, A. J., & Walberg, H. J. (1992a). A process model of mathematics achievement and attitude. *Journal for Research in Mathematics Education, 23*, 306-328.

Reynolds, A. J., & Walberg, H. J. (1992b). A structural model of science achievement and attitude: An extension to high school. *Journal of Educational Psychology, 84*, 371-394.

Rindskopf, D. (1998). Statistical methods for real world evaluation. In A. Reynolds & H. Walberg (Eds.), *Evaluation research for educational productivity* (Vol. 7). Greenwich, CT: JAI Press.

Rindskopf, D. (1992). The importance of theory in selection modeling: Incorrect assumptions mean biased results. In H. Chen & P. H. Rossi (Eds.), *Using theory to improve program & Policy evaluations*. Westport, CT: Greenwood Press.

Robbin, A. (1981). Technical guidelines for preparing and documenting data. In R. F. Boruch, et al. (Eds.), *Reanalyzing program evaluations: Policies and practices for secondary analysis of social and educational programs* (pp. 84-143). San Francisco, CA: Jossey-Bass, Inc.

Rumberger, R. W. (1995). Dropping out of middle school: A multilevel analysis of students and schools. *American Educational Research Journal, 32*(3), 583-625.

Slavin, R.E. (1990). Achievement effects of ability grouping in secondary schools: A best evidence synthesis. *Review of Educational Research, 60*, 471-499.

Temple, J. A. & Reynolds, A. J. (1997). *School mobility and achievement: Longitudinal results from a large urban cohort.* Department of Economics, Northern Illinois University, mimeo.

Tuijnman, A.C. (1994). Selection bias in educational research. In T. Husen & T. Neville (Eds.), *The international encyclopedia of education* (2nd ed., Vol. 4, pp. 5379-5385).

U.S. Department of Education. (1997, April). *Focus on NAEP, National Center for Education Statistics, 2*, 1-6. (http://www.ed.gov/NCES/pubs97/97045.html).

Wang, J. (1997). Demystification of hierarchical linear model using SAS PROC MIXED. *Journal of Experimental Education*, forthcoming.

Winship, C., & Mare, R. D. (1992). Models of sample selection bias. *Annual Review of Sociology, 18*, 327-350.

Wolfe, B., Haveman, R. & An, C. (1996). The "Window Problem" in studies of children's attainments: A methodological exploration. *Journal of the American Statistical Association, 91*, 970-982.

Wood, D., Halfon, N., & Scarlata, D. (1993). Impact of family relocation on children's growth, development, school function, and behavior. *Journal of the American Medical Association, 270*, 1334-1338.

Wu, P., & Campbell, D. T. (1996). Extending latent variable LISREL analyses of the 1969 Westinghouse Head Start evaluation to blacks and full year whites. *Evaluation and Program Planning, 19*, 183-191.

Chapter 12

BENEFIT-COST ANALYSIS AND RELATED TECHNIQUES

W. Steven Barnett

Benefit-cost analysis tends to be seen as an exotic approach to educational research and evaluation. Few educational evaluators and even fewer educators are familiar with benefit-cost analysis, and even a casual review of educational research journals is sufficient to show that it is relatively rarely employed. Yet, benefit-analysis is essentially a formalization of the approach to rational decision making that we all apply in daily life: we weigh the advantages (benefits) and disadvantages (costs) of the available alternatives and choose the one with the greatest net advantages. Thus, it is troubling that important educational decisions are made without the aid of formal benefit-cost analysis, and significant improvements in decision making might be facilitated by its more frequent use. This chapter seeks to encourage the increased use of benefit-cost analysis by providing an introduction to the use of benefit-cost analysis in educational evaluation that argues for its usefulness, practicality, and appropriateness.

Benefit-cost analysis encompasses a number of related methods of economic analysis, two of which are considered here—cost analysis, and cost-effectiveness analysis. Cost analysis is simply accounting for the resources required to produce a program or policy. Cost-effectiveness analysis (CEA) relates costs to ordinary

Advances in Educational Productivity, Volume 7, pages 241-261.
Copyright © 1998 by JAI Press Inc.
All rights of reproduction in any form reserved.
ISBN: 0-7623-0253-4

measures of educational outcomes. CEA is especially useful when one wishes to compare alternative approaches to achieving a single educational objective, for example, alternative approaches to increasing reading ability or reducing high school dropout. As will be seen, it becomes more difficult to rely on CEA in making decisions as the number of important educational outcomes to be considered increases and benefit-cost analysis (BCA) becomes more appropriate.

COST ANALYSIS

At its most basic level, cost analysis is the process of gathering and organizing information about the resources required to implement a program. It seeks to identify and estimate the value of all resources used. The standard approach to cost analysis has been described by Levin (1983) as the construction of an "ingredients" model or recipe. The first step is to construct a list of all the ingredients and the amounts of each that are needed to produce the desired program or implement a policy. The second is to determine the unit cost of each ingredient. When the costs of all ingredients have been established, the sum of the costs of all ingredients (in the required amounts) is the estimated cost of the program. This cost may be calculated on a per pupil basis, or it may be calculated for a classroom, school, community, country, or other organizational or political unit and then divided by the number of students to yield cost per pupil.

From an economic perspective, the cost of any ingredient is its "opportunity cost." The opportunity cost is the what must be given up in order to obtain something. In perfectly competitive markets, the opportunity cost of an ingredient equals its price, and in many cases the price of an ingredient provides a good approximation. However, in the public sector and in countries where markets may work very imperfectly or are not used at all, prices may not be a good guide to opportunity costs. In some instances, there may be no prices for an ingredient. For example, if there is substantial unemployment among teachers and the government sets the pay scale for teachers, teacher salaries may not be very indicative of their true cost to society. Indeed, the opportunity cost of hiring additional teachers who are otherwise unemployed may be zero. Alternatively, when resources are donated, as when volunteers provide tutoring or assist teachers in a classroom at no charge, the true cost of this volunteer labor is not zero but the value of the leisure or other activities that the volunteers give up.

When working with cost per pupil, it should be recognized that error in the estimated number of pupils can be as important a source of inaccuracy as errors in the estimated amounts of resources or unit cost of resources. This is particularly true when the cost for an organization such as a nation, city, or even a school is the starting point as one frequently encounters a tendency towards optimism regarding the numbers of student who can be served. In addition, it is often the case that official reports of the numbers of students served are upwardly biased to create a favorable impression or because programs are reimbursed based on the numbers of children served. For example, a careful count of how many students attend school and how many hours each attends school each week may yield a lower number of full-time equivalent students than official attendance data. Similarly, a survey of the population may reveal many fewer participants in a nonformal education program than reported by the program staff. Moreover, if completion of a certain amount of education is required before the program can be considered successfully delivered (e.g., a year of early intervention, a grade of school, a one-semester course, or graduation from high school or college) and turnover is high, there may be very large differences between the numbers of students ever enrolled in a year, the number enrolled at a given time, and the number to whom the program really has been delivered. In some cases only the last is relevant for calculating cost per pupil.

The importance of cost analysis for educational policy making and practice is obvious. How can one identify the best policy or practice without knowing the costs of the alternatives? Yet, the norm in educational research and evaluation is to ignore cost as is evident in the large numbers of research and evaluation studies that recommend one educational practice or another based on outcomes, but present no information about costs. As this chapter's interest in cost analysis is limited to its use in BCA and CEA, little more will be said about cost analysis per se here. Readers interested in more information on cost analysis are referred to Levin's (1983) text and a volume on cost analysis and educational decision making edited by Barnett and Walberg (1994).

COST-EFFECTIVENESS ANALYSIS

Cost-effectiveness analysis (CEA) is used to compare the costs and outcomes of alternative methods seeking to achieve the same goal or goals. When there is a single outcome of interest, the results of a CEA can be summarized in cost-effectiveness ratios computed either as effects divided by costs (outcomes per dollar) or costs divided by effects (dollars per unit of outcome). The alternative with the highest ratio of outcomes per dollar (or lowest ratio of dollars per outcome) is the most efficient, as it provides the greatest gain per dollar invested.

A simple example of CEA is presented in Table 12.1, which presents the results of a comparison of two hypothetical approaches to improving reading achievement. Reading Discovery increases the reading achievement scores of children over a one year period by five months (in grade equivalents). Hooked on Flash Cards increases reading achievement by two months. Comparing only outcomes, Reading Discovery seems to be superior. However, Reading Discovery is much more expensive as it costs $2000 per pupil for a year while a year of Hooked on Flash Cards costs $300 per pupil. The cost-effectiveness ratios reveal that Hooked

Table 12.1.   Hypothetical Cost-Effectiveness
Results for Reading Interventions

Approach	Effects	Cost	C-E Ratio
Reading Discovery	5 months	$2,000	$400
Hooked on Flash Cards	2 months	$300	$150

on Flash Cards is more efficient at $150 per month gain in reading score compared to $400 per month gain for Reading Discovery.

Although Hooked on Flash Cards has the most favorable cost-effectiveness ratio, this does not necessarily mean that it should be adopted over Reading Discovery. Recall that Reading Discovery produces two and a half times the gain in reading scores. If the amount of time spent on Hooked on Flash Cards could be increased two to three times with proportional increases in results, then reading scores could be improved by four to six months in grade equivalents at a cost of $600 to $900. In this case the choice is obvious. However, it might be that Hooked on Flash Cards is not easily expanded, that increased time spent in the program yields little additional gains or that doubling or tripling the time for Hooked on Flash Cards could only be accomplished by drastically reducing time spent on other subjects. In this case the decision is less clear. If only one program can be adopted, the school's leadership will have to decide whether the additional three months gain in reading scores from Reading Discovery is worth the additional cost of $1700 per pupil ($2000 minus $300). If the budget allows, it also might be possible to adopt both programs. Whether this makes sense depends on whether the two programs together yield seven months gain, or something less, and how much the added gain is valued by the leadership.

An Example: Comparing the Cost-Effectiveness of Four Primary School Reforms

A more complex example is provided by a CEA of four educational reforms promoted as ways to improve student achievement in the primary grades (Levin, 1988; Levin, Glass, & Meister, 1984). The approaches examined are: a longer school day, computer assisted instruction (CAI), cross-age tutoring (by peers and adults), and reduced class size. Levin and colleagues produced the analysis by adding cost estimates to measures of effectiveness obtained from reports of previously conducted studies. Their results are presented in Table 12.2. Cost per child is twice as high as cost per child per subject for each reform except class size for which cost is three times as high as the cost per subject (based on the assumption that reading and math account for two-thirds of class time).

Overall, the results in Table 12.2 show that the annual costs of most of the reforms are relatively low, but the effects tend to be modest, as well. CAI has a

Table 12.2. Costs and Effects
(months of grade equivalents) of Four Primary Grade Reforms

Type of Reform	Cost per Student	Cost per Subject	Math Gain	Reading Gain
Longer Day	$122	$61	0.3	0.7
CAI	$238	$119	1.2	2.3
Peer Tutors	$424	$212	9.7	4.8
Adult Tutors	$1,654	$827	6.7	3.8
Reduced Class from 35 to 20	$402	$201	2.2	1.1

Source: Levin (1988).

noticeable impact on achievement as does a reduction in class size from 35 to 20 students, but with different relative impacts on reading and math. Moreover, the reduction in class size is much more costly than CAI. Cross-age peer tutoring generates much larger gains than the other reforms, at about the same cost per subject as very large reductions in class size. The adult tutoring component of cross-age tutoring produced the next largest gains, but is much more costly than the other options.

The results are presented in a ratio form in Table 12.3, courtesy of Levin (1988). This makes the alternatives easier to compare by giving the projected annual cost of producing one month gain in student achievement in the two subject areas for each approach. In terms of the cost-effectiveness ratios, peer cross-age tutoring is clearly the best alternative. Assuming that math and reading scores are considered approximately equally important, CAI ranks next. Depending on the relative importance assigned to reading or math, either the longer school day or reduced class size follows. The adult component of cross-age tutoring ranks last due to its high cost.

Given this information, what should a school do to improve student achievement? Choose the alternative with the best cost-effectiveness ratio? The answer depends on the school's situation and what happens to the results when the scale of a reform approach is changed from what was tested. For example, assume that a school system has $500 per pupil to spend on primary school reform. In that case, peer tutoring is the obvious choice. However, suppose a school has only $250 per pupil for reform. If it adopts the peer tutoring program, it can fund the participation for only about half the students. If it adopts the CAI program, it will be able to fund participation of all the students. One response might be operate the peer tutoring program in half of the grade levels so that all students would participate at some time. Alternatively, the district might be able to provide all students with math tutoring only (which showed the greatest gains) or reduce the number of tutoring sessions by half from the model tested. However, the school runs the risk that there are spillovers from reading tutoring to math or that half the tutoring is less than half as effective.

Table 12.3.   Cost-Effectiveness Ratios for Four Primary Grade Reforms

Type of Reform	Cost/Month Math Gain	Cost/Month Reading Gain
Longer Day	$203	$87
CAI	$100	$52
Peer Tutors	$22	$44
Adult Tutors	$123	$218
Reduce Class Size 35 to 20	$91	$183

Source: Levin (1988).

Another school system might be in the situation of deciding whether or not to raise money for school reform, and, if so, how much should be raised. If the leadership believes that a 10 month gain in math and five month gain in reading is worth at least $424 per child, then it ought to seek funding to implement the peer tutoring program. If it does not believe that the gains are worth $424 dollars per child, it probably should not pursue any of the alternatives. If it decides to implement peer tutoring, it should not necessarily stop there. The system should then consider the next highest ranking option. If it is agreed that spending roughly $240 per child for another one month increase in math scores and two month increase in reading scores is worthwhile, then the CAI program should be funded as well. This examination of options should continue until it is determined that the next most efficient option is undesirable or that the additional funds cannot be raised.

Obviously, even this example has been simplified. In practice, decisions about funding, policy, and practice are complex, and extrapolation from research studies to any specific school's situation must be undertaken cautiously. The estimated costs and effects for each program depend on the scale and level of implementation as well as the context. As mentioned earlier, peer tutoring might not work as well if the number of sessions was cut in half or tutoring was in math only. Farkas (1996) has studied the effects of tutoring using paraprofessionals. He finds roughly the same size effects, but the cost is only about one-quarter the cost of tutoring with teachers. The cost per month of this adult tutoring is about the same as the cost of peer tutoring. In this case, a choice between peer tutoring and adult tutoring might be based on the school leadership's assessment of its ability to faithfully implement the model as described in the research reports.

Finally, the estimates of costs, effects, and cost-effectiveness ratios presented in Tables 12.1-12.3 are all point estimates. This is fairly typical of the presentation of CEA results. However, there is always some uncertainty surrounding such estimates. One way to deal with that uncertainty is to conduct a sensitivity analysis in which assumptions about costs and effects are varied to produce a range of estimated costs, effects, and cost-effectiveness ratios. The other way is to report confidence intervals. In education, confidence intervals may be available only for the estimated effects because cost per pupil is estimated as total cost for a program at a single site (school, district, state, etc.) divided by the number of students. In this case there is no real information on variation in cost. In other cases, there is a cost figure for each pupil, and both the numerator and denominator of the cost-effectiveness ratio contain stochastic elements. With small sample sizes (which are common), the distribution of the ratio estimator is likely to be skewed, and methods of confidence interval estimation that account for this are recommended (Chaudhary & Stearns, 1996).

BENEFIT-COST ANALYSIS

Benefit-cost analysis (BCA) can be viewed as an extension of cost-effectiveness analysis in which dollar values are attached to effects in the same way that dollar values are attached to ingredients in CEA. Similarly, CEA may be viewed as an incomplete BCA. For example, when analyzing school reform models such as Success for All and Accelerated Schools that are reported to reduce grade repetition and increase achievement one could estimate the cost-savings from reduced grade repetition and subtract this from cost prior to computing cost-effectiveness ratios for reading, math, and other subjects. Keep this overlap in mind while reading this section as much of what is said here regarding methods applies to complex CEA's as well as to BCA.

The primary advantage of BCA over CEA in educational research and evaluation should be clear from even the simple examples of CEA used above. Typically, educational services have multiple outcomes so that each alternative policy or program has multiple cost-effectiveness ratios (and some outcomes may occur for one option and not another), and the use of cost-effectiveness ratios does not always eliminate the need for judgments about the value of effects relative to their costs even when there is only a single outcome of interest. By taking the further step of explicitly valuing effects, BCA provides a clearer guide to the economic efficiency (what we get out for what we put in) of alternative programs and policies. Of course, these advantages are not obtained without cost. Much more effort tends to be required to produce a BCA than a CEA, and the data requirements can pose difficulties as well. One purpose of this chapter is to demonstrate both the possibilities and difficulties of BCA.

Returns to Schooling

Despite the lack of benefit-cost analysis in education generally, there is one area of educational evaluation in which hundreds of benefit-cost analyses have been conducted. This is the estimation of the returns to the quantity of schooling and formal and informal training for workers. Reviews of this literature have been provided by Cohn and Addison (in press), Hough (1994), and Psacharopoulos (1994). Cohn and Addison (in press) provide an excellent review of recent studies of the returns to academic education and vocational training in Europe, the United States, Canada, and Australia. They find that rates of return to schooling vary greatly within and across countries, but overall studies tend to find that education is a sound private and public investment at primary, secondary, and post-secondary levels. These positive results are found despite the fact that the overwhelming majority of studies considered only the benefits from increased earnings. Data on earnings and education are rather common. Data that can be used to estimate the effects of schooling on other outcomes of education is rare, and one then must face the problem of estimating the dollar value of those outcomes. Findings regarding the returns to vocational training tend to be more pessimistic, except for employer-sponsored on-the-job training, though further research is called for based on controversies over the adequacies of existing research. Focusing on government training programs for the poor, Friedlander, Greenberg, and Robbins (1997) reach more positive conclusions about rates of return, especially for women. Also noteworthy is the hotly debated literature on the returns to the *quality* of education, though few studies in this field can be said to have reached the stage of true BCA (reviews from a variety of perspectives are provided by Burtless, 1996). Ludwig and Bassi (1997) indicate that model misspecification may account for past negative findings and produce new results indicating that returns to quality are substantial.

BCA Methodology

To demonstrate the practicality of more comprehensive approaches to BCA while providing concrete illustrations of procedures, this section explains BCA methodology with the aid of a well-known, and admittedly unusual, study. The Perry Preschool Project is a longitudinal study of the effects of preschool education provided to three- and four-year-old African-American children whose families were in poverty (Schweinhart, Barnes, Weikart, Barnett, & Epstein, 1993). The 123 subjects of the study were randomly assigned to preschool or control groups in five waves in the mid-1960s. They were followed-up annually through age 10 and, thereafter, at ages 14, 15, 19, and 27.

The primary goal of this educational intervention was to improve the children's school success, but it was conducted with a broad view of children's development and had effects as complex as will be encountered in any education study. The BCA employed data collected through age 27.

The process of conducting a BCA can be described as a series of steps or categories of activities (Thompson, 1980). Depending on the level of detail desired, a greater or smaller number of steps might be listed. However, the following eight steps capture the basics and are explained below:

1. Define the scope of the analysis.
2. Estimate cost.

3. Estimate effects.
4. Estimate benefits.
5. Adjust for inflation.
6. Calculate the return.
7. Describe distributional consequences.
8. Conduct sensitivity analysis.

Define the Scope of the Analysis

Like all types of evaluation, BCA is comparative. The first step is to define the alternatives to be compared and the boundaries of the evaluation. The definition of alternatives tends to be easy when the comparison is between one or more specific programs. It can be more difficult when a single policy or program is under evaluation, and the state of the world without the policy or program must be defined. In setting the boundaries of the analysis, the evaluator must decide whose costs and benefits will be estimated and what program outcomes will be investigated.

The BCA of the Perry Preschool Project was designed to inform public policy regarding preschool education for economically disadvantaged children nationally. Thus, scope of the study was defined from a national perspective. It was decided to examine the total costs and benefits to society as a whole and to divide these into those accruing to preschool participants (and their families) and those accruing to the general public. Theory and previous research findings provided the basis for an outline of potential developmental, academic, social and economic effects (positive and negative) of the preschool program. The analysis included as many of these as possible, but data availability limited the possibilities. For example, one might expect (small) spillover effects on the nonattending siblings of children who attended the preschool program. However, effects on siblings could not be estimated as data were not collected for siblings.

Estimate Cost

Following the ingredients approach to cost analysis, this step began with a detailed description of the program and a list of all the resources required. The Perry Preschool program was operated by a public school from October to June and consisted of a two and a half hour per day classroom program which met five days per week and one and a half hour weekly home visits by the teachers. Children attended the program at ages three and four, except for a small group of children the first year who attended only at age four. Detailed data on staffing, facilities, and other resources were obtained from project and school district records and were used to estimate the full cost of the program to taxpayers, including the capital costs of buildings and land. In addition to the direct cost of instructional services, children attending the preschool program were allocated a share of district-and school-level administrative (and support) services costs equal to the per pupil average for all children in the district. Policymakers sometimes want to assume that new educational programs can be added to a school with no (or very little) increase in administrative and support costs. This should be resisted as added programs tend to require at least as much administrative support, supervision, in-service training, evaluation, and record keeping as existing programs. The program was estimated to cost approximately $7200 per child per year in 1992 dollars.

Estimate Effects

Every BCA is built upon an evaluation of program effects, and the validity and usefulness of a BCA is limited by the strength of its underlying evaluation. The Perry Preschool program's effects were estimated from a prospective longitudinal randomized trial that provides a relatively strong basis for valid estimates. An extremely wide range of child and family pre- and post-intervention measures were collected over 25 years of data with remarkably little attrition. Such randomized trials with pre- and post-test data are rare in educational research and evaluation, despite the difficulties of producing valid estimates of the effects of educational programs with weaker designs (Cook & Campbell, 1979).

The evaluation of the Perry Preschool program found a chain of lasting effects stretching from preschool to adulthood. Eight key findings are reported in Table 12.4 in the form of comparisons between the preschool and control groups. As these are reported for illustrative purposes, they will not be discussed in detail. However, information on the estimated effects of the program and the analytical methods used to estimate effects on these and other outcome measures has been extensively reported elsewhere (Schweinhart et al., 1993; Berrueta-Clement, Schweinhart, Barnett, Epstein, & Weikart, 1984; Schweinhart & Weikart, 1980).

Estimate Benefits

The estimation of benefits (the economic or monetary value of effects) is the core of BCA and marks the point at which it diverges from CEA. This is the most difficult step in BCA because many of the effects of education can present problems for monetary valuation. For example, if the Perry Preschool study had ended its follow-up prior to high school, it would have provided evidence of effects on early IQ (that fade away as the children grow older) and persistent effects on achievement test scores. Although these effects clearly are beneficial, it is difficult to judge how much they are worth. Links might be made between achievement test scores and earnings, but any estimates of dollar values would be highly imprecise and would capture only part of the benefits. Two other effects were found for which it was rather easy to estimate an economic value. First, the program itself provides child care as do all classroom-based educational programs. Educators tend to forget that schooling provides this benefit to working parents regardless of

Table 12.4. Selected Effects of the Perry Preschool Program

Outcome Measure	Preschool	Control	N
IQ at Age 6	91	86	120
Achievement at 14	122	95	95
Classified EMI	15%	34%	123
High School Grad.	66%	45%	123
Average or Better Literacy at Age 19	61%	31%	109
Arrests by Age 27	2.3	4.6	123
Ever on Welfare	59%	80%	123
Monthly Earnings at Age 27	$1,219	$ 766	116

Notes: All differences between preschool and control groups are statistically significant at $P < .05$. Achievement at age 14 was measured by California Achievement Test (Level 4) raw score.

its educational impacts. Second the preschool program reduced placements in special education of children classified as educable mentally impaired and lowered the number of years of special education children received. This reduced the costs of education by an amount that is easily calculated. Other preschool programs have been found to reduce the amount of grade repetition, another effect with an easily calculated monetary (Barnett, 1995). Even this should not be made to sound too simple; without follow-up through the end of high school, some projection would be necessary of special education and grade repetition beyond the follow-up period.

Estimation of benefits from the Perry Preschool study was much more feasible than in many studies because follow-up continued through the end of high school. Data are available from the census and large scale national surveys linking number of years of schooling and high school graduation to important adult outcomes such as employment and earnings, reliance on welfare, child-bearing and family formation, and crime. By using estimates from these data sets to link effects on educational attainment to these other adult outcomes, it is possible to estimate key long-term economic benefits. Thus, benefit-cost analysis of educational programs and policies is much more likely to be feasible when effects on educational attainment are expected and can be measured. Increased educational attainment might be expected from any educational intervention seeking to improve achievement and school progress (i.e., reducing grade repetition and special education needs), especially if low-income or poorly performing students are targeted.

As Table 12.4 indicates, the Perry Preschool study's advantages for benefit estimation go significantly beyond data on educational attainment. The study provides direct estimates of effects on such important adult outcomes of education as crime, employment and earnings, and reliance on welfare. Complete results of the benefit-cost analysis are presented in Table 12.5, with benefit estimates provided in 5 categories child care, education, employment, crime, and welfare. The child

Table 12.5.   Present Value of Costs and
Benefits of the Perry Preschool Program

Costs and Benefits	(A + B) Society	(A) Participants	(B) Taxpayers
Preschool Cost	−$12,356	$0	−$12,356
Measured Benefits to age 27			
Child Care	738	738	0
Education	6,287	0	6,287
Employment	14,498	10,269	4,229
Crime	49,044	0	49,044
Welfare	219	−2,193	2,412
Benefit Subtotal	$70,876	$8,814	$61,972
Projected Benefits beyond age 27			
Earnings	15,833	11,215	4,618
Crime	21,337	0	21,337
Welfare	46	−460	506
Total Benefits	$108,002	$19,569	$88,433
Net Present Value	$95,646	$19,569	$76,077

care benefits are minor as the program provided only a small number of hours of services and the value of custodial care to parents is low (about $1.50 an hour). The education benefits were estimated as the difference in costs (year-by-year) for educating the preschool and control groups (cost savings from reduced special education needs were partially offset by cost increases from students continuing in school longer).

Three benefits presented more complex problems for estimation, including the need to make projections beyond the last follow-up for data collection—employment, crime, and welfare. Many benefit-cost analyses of education policy and programs must face the problem of projecting benefits beyond the last period of data collection because education produces lifelong benefits. Benefits through age 27 are presented in the top part of Table 12.5. Benefit estimates for ages 28 through 65 are presented in the bottom part of Table 12.5 and are based on projections into the future that require assumptions not necessary to produce estimates up to age 27. A brief description of the estimation process for each benefit is informative (for further details see Barnett, 1993).

Employment benefits through age 27 were estimated as gains in earnings plus fringe benefits. Earnings were estimated from employment and earnings histories obtained at ages 19 and 27. Fringe benefits were not reported, so these were estimated using data from other studies relating the value of fringe benefits to earnings. National data relating earnings by age over the lifecycle to educational attainment made it possible to project earnings beyond age 27 from educational attainment by age 27. Procedures for such projections have been well-developed for years (Miller, 1960; U.S. Bureau of the Census, 1983). Projections were made using national earnings data for blacks by age, sex, and five educational attainment levels: less than high school completion, high school completion, some college, four years of college, and more than four years. Earnings estimates were adjusted at each age by the probability of survival as estimated from national data for blacks by age and sex. Productivity growth was assumed to be zero over this time period, so no adjustment was made for growth over time. Again, national data linking fringe benefits to earnings were used to estimate fringe benefits.

Although well-developed and highly defensible, this approach to projecting benefits from employment was only partially satisfactory. First, it captures only the monetary benefits from employment. It does not capture any benefits from increased job satisfaction, better working conditions, or the contributions of employment to self-esteem and status. Second, the preschool program resulted in gains in educational attainment only for women. The men who had attended pre-school had higher achievement test scores (they learned more), but did not differ in how much schooling they completed. Data through age 27 demonstrated that the men nevertheless had higher earnings (because of their higher skill levels), but the basis for projecting future earnings from skill levels rather than educational attainment was judged too weak to be pursued. Thus, the analysis very likely underestimates earnings benefits.

Benefits from reduced crime were estimated in the form of reductions in two types of costs—costs to victims (property loss, lost work time, and pain and suffering) and costs to the criminal justice system. Reductions in crime costs through age 27 were estimated from the participants' arrest and criminal justice system histories, combined with national data on the number of crimes per arrest, costs to victims, and the costs of police, court, and prison services. Data on victim costs a study etimating costs to victims by type of crime based on jury awards to people who had suffered injuries (Cohen, 1988). Reductions in crime costs beyond age 27 were projected based on arrest data through age 27 under the assumption that criminal activity by each person from age 28 to age 65 would be proportional to criminal activity prior to age 27. Data on number of arrests by age and sex for each type of crime were obtained from the *Uniform Crime Reports* (Federal Bureau of Investigation, 1992).

Benefits through age 27 from reductions in dependence on welfare were simply estimated by comparing the amounts of welfare that individuals in the two groups had received. The welfare payments themselves are a shift in resources from some members of society to others, what economists call a "transfer payment." These payments affect the distribution of income, but do not reduce the total resources society has available. Thus, from the perspective of society as a whole the real cost of welfare dependency (apart from psychological and status effects on the recipients not measured in this analysis) is the administrative cost, which is about 10 percent of the amounts of welfare payments. Thus, the benefit to society was estimated to equal 10 percent of the amount by which payments were reduced. Lacking national data relating welfare program participation at each age to education, future welfare payments were projected using data from the Panel Study on Income Dynamics (Ellwood, 1986) to forecast the probability of exiting welfare each year given the prior number of years on welfare.

Adjust for Inflation

Every BCA that deals with costs and benefits over more than one year has to make adjustments for differences in the value of money over time due to inflation. Even low to moderate rates of inflation such as the United States has experienced over the last few years can have substantial effects on dollar values as the effects of inflation are compounded from one year to the next (just like compound interest). Inflation distorts comparisons over several years, and because in educational evaluations it is typical that costs are incurred up front while benefits are received later, the failure to adjust for inflation tends to overstate net benefits. Similarly, cost comparisons among two or more programs can be highly misleading, if the cost estimates are from different years and no adjustments are made for inflation. Outside the United States, inflation rates are sometimes very high, 50 percent, 100 percent, and higher. In one recent year, Russia had an annual inflation rate of 2500 percent. When inflation rates are very high, even differences of a few months in the timing of costs and benefits must be taken into account to accurately assess the value costs and benefits. Adjustments for inflation are made by using a price index to convert current dollars from various years into constant, or real, dollars in a single year. All of the estimates in Table 12.5 were adjusted for inflation by converting dollar values from each year into 1992 dollars using an implicit price deflator for Gross Domestic Product (GDP; U.S. Bureau of Economic Analysis, 1993). Implicit GDP deflators are less well-known to the general public than the consumer price index (CPI), but more appropriate for use in BCA of education programs and policy. Recent work on price indices by the government has increased the number of different types of deflators available for use in the United States (Triplett, 1992; Young, 1992).

Calculate the Return

Even in the absence of inflation, a dollar today is more valuable than a dollar next year or some other time in the future because there is an opportunity cost of waiting for the future dollar. If we have the resources today, they can be consumed for our present enjoyment or invested to yield or produce additional resources. We give up current consumption or investment gains, when we must wait for a program with current costs to produce future benefits. There are several alternative ways to take this opportunity cost into account. The most commonly used are the calculation of "discounted present value" and the "internal rate of return." Present value is the value of all resources (in constant dollars) from any year discounted at a rate that equals the annual opportunity cost (that is, the rate at which we are willing to trade dollars in one year for dollars in the next year). After discounting, the present values of all costs and benefits are summed to calculate net present value. The internal rate of return is calculated as the discount rate at which the sum of the present values of costs and benefits equals zero. In other words, the internal rate of return is the highest discount rate at which net benefits are nonnegative.

The economic desirability of policies and programs can be judged from their net present value or internal rate of return, and alternative programs can be ranked using these methods. A positive net present value or high internal rate of return is a clear indicator that a program or policy can make a positive economic contribution to society and may be a sound public investment. However, economists disagree about the "correct" discount rate to use in calculating present value (and, therefore, also disagree about what is a sufficiently "high" rate of return), and no single rate is appropriate for every analysis. This is one reason that some economist prefer to calculate the internal rate of return. Alternatively, the analyst can calculate net present value using a range of discount rates. Most economists would be comfortable with a range of real rates (sometimes nominal rates which include an adjustment for inflation are used) from zero to 7 percent. There is an extensive literature on discount rates that can be consulted prior to selecting a discount rate or making judgments about an internal rate of return (Just, Hueth, Smith, & Schmitz, 1982; Kolb & Sheraga, 1990; Lind, 1982; Mikesell, 1977; Mishan, 1976; Thompson, 1980).

Describe Distributional Consequences

Policies and programs differ in *who gains* and *who loses* as well as in their economic return to society as a whole. Net present value can be equal for projects that benefit everyone equally, benefit the poor more than the rich, or benefit the rich at the expense of the poor. Obviously, people and their political representatives are not indifferent about these distributional differences. If a BCA is to be useful, it must describe the distribution of costs and benefits meaningfully. Usually a relatively small additional effort is required, as most of the work has been done in estimating costs and benefits. All that is required is the allocation of costs and benefits to different sectors of society. Exactly how this is done depends on the relevant concerns in each case.

In the analysis of the Perry Preschool program's impact, costs and benefits were divided between study participants, who were very low income minority families, and the general public. Note that this is a conceptual distinction as the participating families are also part of the general public. This distributional description was considered useful for two reasons. First, the goal of the program was to significantly improve the lives of children growing up in poverty, and it was important to show the extent to which the program accomplished this goal. Second, the program generated substantial spillovers to the general public (also called public benefits) in addition to the private benefits produced for the participants. Thus, the distributional analysis provides information about how much of a return the general public might expect on their investment beyond the accomplishment of the program's more altruistic objectives.

The distribution of costs and benefits for the Perry Preschool program is described in the last two columns of Table 12.5. The estimated economic value of benefits to participants and their families is quite small until the children become adults and the preschool group earns higher incomes. Even then the gains to the participants are partly offset by reductions in welfare payments received. Overall, the gains to the participants alone are estimated to more than cover the cost of the program. Net present value would be positive without any benefits to the general public. Moreover, participants' benefits were underestimated because the dollar value of such intangible benefits as obtaining more satisfaction from schooling and increased status due to more regular and higher paying employment could not be estimated. The estimated monetary gains to the general public are quite large, consistent with other research indicating that spillover effects from education generally are substantial. These results are of great interest in a policymaking context because they indicate self-interest as well as altruism would be served by public funding of high quality preschool education for children in poverty.

Conduct Sensitivity Analysis

Every BCA requires some assumptions to produce estimates of costs, benefits, and net present value. Uncertainty about these assumptions may be due to theoretical or empirical disagreements over the correct assumption or simply because predictions about the future are required. These assumptions may or may not prove to be true, and variations in the assumptions can alter the results of the BCA. Sensitivity analysis is used to identify critical assumptions and to explore the effects of reasonable variations in assumptions on the results. One of the most important uses of sensitivity analysis is to indicate conditions that must hold in the future for a proposed policy or program to be a worthwhile investment. If such conditions are identified, steps may be taken to ensure that these conditions hold when the policy or program is begun.

As noted earlier, one assumption that is always a matter of contention is the appropriate discount rate. In the Perry Preschool CBA, real discount rates of 0, 3, 5 and 7 percent were used to produce separate analyses that encompassed the reasonable range of disagreement about real discount rates. Net present value was found to be positive at all of these rates, so that the choice of a particular discount rate was not critical in this instance. Further analyses with even higher discount rates found that the real internal rate of return exceeded 11 percent.

The bottom line on the results in Table 12.5 is that the estimated net present value of the Perry Preschool program to society was found to be positive under a wide range of variations in assumptions including the complete exclusion of many of the estimated benefits. When consideration is given to the potential value of benefits not included in the analysis (for example: the intrinsic value of increased literacy and its value in leisure, family life, and citizenship; increased marriage rates and decreased abortion rates for the girls; increased status associated with greater economic success and educational attainment) the results appear to be even more robust. Had the benefit estimates not been so large already, it might have been useful for the sensitivity analysis to recalculate net present value using a variety of assumptions for the annual value of benefits omitted from the analysis.

LIMITATIONS OF BCA

BCA and the related approach of CEA both have practical limitations. Neither analysis can be any better than the underlying descriptions of resource requirements and evaluations of program effects that provide the basic materials for their analyses. BCA has limitations that CEA does not have due to the estimation of the monetary value of program effects. The most obvious is that without very long-term outcome measures it is extremely difficult to estimate the monetary value of the effects of most educational policies and programs. Much of the research on returns to quantity and quality of education has tended to omit measures of direct educational effects altogether and skip directly to measures of earnings. Estimates by Haveman and Wolfe (1984) suggest that analyses limited to effects on earnings capture only about half the value of long-term benefits from education. The results of the Perry Preschool BCA are consistent with this, but also indicate the substantial data requirements for more complete analyses. One possible remedy for this problem is research on the nonearnings effects of educational attainment and achievement that could then be linked to a large number of studies of educational outcomes. Of course, this assumes that education's benefits on such outcomes as crime, fertility, health, and leisure activities are the result of its academic outcomes rather than socialization outcomes, an assumption that deserves further study. For the present, evaluators must be aware that because it is much easier to fully estimate costs than benefits, BCAs tend to underestimate the economic value of educational policies and programs.

Several responses to the difficulty of measuring benefits can be proposed. The simplest is to collect more extensive and better data, but this is also the most costly. Another is to use survey methods to elicit monetary values for effects of education for which prices are not found in the market (test score gains or improvements in daily living skills, for example). This approach has been used in education with limited success (Escobar, Barnett, & Keith, 1988), and its potential usefulness in BCA generally is controversial (Hausman, 1993). A third response is to conduct a hybrid CEA/BCA in which one or more program effects are valued, but other highly important effects are not. By estimating the dollar value of effects where this is possible and subtracting their present value from the present value of costs, the analyst can reduce uncertainty regarding how much value must be attached to the remaining effects in order for the policy or program to be judged economically sound or the best among the alternatives. The results of such a hybrid analysis can be presented in a CEA format.

BCA is sometimes criticized for being concerned with only monetary values and not human values. The response to such criticism is that the use of money to measure value need not limit the analysis to economic phenomena (though the bias in this direction in past estimates of the returns to education is overwhelming). Moreover, the money values used represent the human values of individual members of the society as expressed in their free choices in the market. When market prices do not accurately represent the values of a society's members, the analyst is expected to estimate shadow prices that better represent those values. From this perspective, economics brings together the values and competing interests of all the persons affected by a policy or program.

There is more than a little appeal to the economic approach to taking values into account. It is a relatively democratic approach that many have found acceptable, and it can be implemented. Other approaches can be conceived, but to this point have not shown much practical success or achieved widespread acceptance. For example, market prices may be viewed as an inappropriate basis for political decisions because they reflect the distribution of income and wealth as well as values. However, the prices that would arise under a more equal distribution are unknown. Others might argue no aggregation of individuals' values was an appropriate basis for determining the best decision.

Even judged from a friendly perspective, the economic approach has some shortcomings. Some values resist inclusion in an analysis by way of prices. Values that are weak and widely dispersed may not have a measurable effect on prices. There may be some gains or losses for which prices are less meaningful. For example, there may be no price for which a person would give up eyesight, physical mobility, or life. On the other hand, people do pay to avoid risks of these losses, and these payments provide estimates of the value of avoiding injury and death. Some economists have mislead the public by writing about estimating the value of a life, when what they really had estimated was the value of a livelihood (i.e., lifetime earnings potential). Nevertheless, it can be useful to estimate the economic value of saving a life (or reducing the risk of death), which is roughly an order of magnitude higher than the estimated value of a livelihood, because public policy may tend to undervalue lives in the absence of such estimates (Viscusi, 1986).

Another difficulty is presented by rights. Rights do not have market prices, but often are central issues in education—rights to due process, to equal education, to be involved in educational decisions about one's children, to a free appropriate education. However, rights can be taken into account by treating them as constraints. For example, an analysis could compare only policies that ensured the rights of every student were protected. Also, BCA can be used to estimate the costs to schools and other education agencies when new rights are introduced. Given, the high costs of special education, studies of the costs and benefits of rights to special education services might be a productive area for research.

Finally, whatever the ethical limitations of economic evaluation as a basis for making decisions, it can be argued that it is prudent to have an accounting of the economic value of costs and benefits of the alternatives. In this view, BCA or CEA need not be the only considerations in making a decision, but the decision should made with at least an awareness of its economic consequences. Currently, this rarely happens because BCA and related methods are rarely employed in educational evaluations.

CONCLUSION

This chapter seeks to interest evaluators in increasing the use BCA and related methods both directly and through the inclusion of economists in multidisiplinary evaluation teams. Where evaluation produces quantitative measures of effectiveness, the move to CEA may require only the addition of a cost analysis. In many cases, educational evaluators could conduct a reasonable CEA using the information presented in a brief guide to cost analysis (e.g., Levin, 1983). In some cases, decisions might be improved if educational administrators simply applied the logic of BCA to the results of an evaluation that provided rough indications of costs as well as measures of effects.

Increased BCA in educational evaluation will be more costly, but the costs of BCA may be small compared to the costs of ill-informed decisions in many situations. The difficulties involved in BCA should not be underestimated. To some extent these difficulties might be reduced by a program of research to develop benefit estimates that can be linked to outcomes commonly measured in educational evaluations. However, the success of such a research program is not guaranteed, and it may be thwarted by complex interactions of person, process, and context that affect the outcomes of educational programs and practice. Studies that seek to directly estimate benefits such as the BCA of the Perry Preschool program impose substantial burdens for data collection (though lesser efforts also result in more limited estimates of effects) and analysis. Larger sample sizes are likely to be required to find a significant difference between costs and benefits than to detect a positive effect. Typically, BCA will require that an economist specializing in BCA join experts in the subject matter and evaluation to plan and conduct an evaluation. Economists who understand education and human development are more likely to provide the insights needed to produce BCA of an educational program or policy. Investments in BCA as part of evaluations designed to inform state and national decision making (for example, about alternative approaches to systemic reform, state and national testing programs, or subsidies for higher education) might have large payoffs by providing a basis for better decisions.

REFERENCES

Barnett, W. S. (1993). *Lives in the balance: Age-27 benefit-cost analysis of the High/Scope Perry Preschool Program.* Ypsilanti, MI: High/Scope Press.

Barnett, W. S. (1995). Long-term effects of early childhood programs on cognitive and school outcomes. *The Future of Children, 5*(3), 25-50.

Barnett, W. S. & Walberg, H. (Eds.) (1994). *Cost-analysis for educational decisions: Methods and examples. Advances in educational productivity* (Vol. 4). Greenwich, CT: JAI Press.

Berrueta-Clement, J. R., Schweinhart, L. J., Barnett, W. S., & Epstein, A.S. (1984). *Changed lives: The effects of the Perry Preschool program on youths through age 19.* Ypsilanti, MI: High/Scope Press.

Burtless, G. (Ed.) (1996). *Does money matter: The effect of school resources on student achievement and adult success.* Washington, DC: Brookings Institution Press.

Chaudhary, M. A., & Stearns, S. C. (1996). Estimating confidence intervals for cost-effectiveness ratios: An example from a randomized trial. *Statistics in Medicine, 15*(13), 1447-1557.

Cohen, M. A. (1988). Pain, suffering, and jury awards: A study of the cost of crime to victims. *Law & Society Review, 22*(3), 537-555.

Cohn, E., & Adison, J. T. (in press). The economic returns to lifelong learning. *Education Economics.*

Cook, T. D., & Campbell, D. T. (1979). *Quasi-experimentation: Design and analysis issues for field settings.* Boston: Houghton Mifflin.

Ellwood, D. (1986). *Targeting the would-be long-term recipient of AFDC: Who should be served?* Princeton, NJ: Mathematica.

Escobar, C. M., Barnett, W. S., & Keith, J. (1988). A contingent valuation approach to measuring the benefits of preschool education. *Educational Evaluation and Policy Analysis, 10*(1), 13-22.

Farkas, G. (1996). *Human capital or cultural capital.* New York: Aldine deGruyter.

Federal Bureau of Investigation. (1992). *Crime in America* (Uniform Crime Reports). Washington, DC: USGPO.

Friedlander, D., Greenberg, D., & Robins, P. K. (1997). Evaluating government training programs for the economically disadvantaged. *Journal of Economic Literature, 35*(4), 1809-1855.

Hausman, J. A. (Ed.) (1993). *Contingent valuation: A critical assessment.* New York: North-Holland.

Haveman, R. H., & Wolfe, B. L. (1984). Schooling and economic well-being: The role of nonmarket effects. *Journal of Human Resources, 14*(3), 377-407.

Hough, J. R. (1994). Educational cost-benefit analysis. *Education Economics, 2*(2), 93-128.

Just, R.E, Hueth, D.L, & Schmitz, A. (1982). *Applied welfare economics and public policy.* Englewood Cliffs, NJ: Prentice Hall.

Kolb, J., & Sheraga, J. (1990). Discounting the benefits and costs of environmental regulations. *Journal of Policy Analysis and Management, 9*(3), 381-390.

Levin, H.M. (1983). *Cost-effectiveness: A primer.* Beverly Hills: Sage.

Levin, H.M. (1988). Cost-effectiveness and educational policy. *Educational Evaluation and Policy Analysis, 10* (1), 51-69.

Levin, H.M., Glass, G.V., & Meister, G.R. (1984). *A cost-effectiveness analysis of four educational interventions.* Stanford, CA: Stanford University, School of Education.

Lind, R. (Ed.) (1982). *Discounting for time and risk in energy policy.* Baltimore: Johns Hopkins University Press.

Ludwig, J., & Bassi, L. J. (1997). *School spending and student achievement: New evidence from longitudinal data.* Paper presented at the APPAM Research Conference, Washington, DC.

Mikesell, R. (1977). *The rate of discount for evaluating public projects.* Washington, DC: American Enterprise Institute.

Miller, H.P. (1960). Annual and lifetime income in relation to education. *American Economic Review, 55,* 834-844.

Mishan, E.J. (1976). *Cost-benefit analysis.* New York: Praeger.

Psacharopoulos, G. (1994). Returns to investment in education: A global update. *World Development*, 22(9), 1325-1343.

Schweinhart, L.J., Barnes, H. V., Weikart, D. P., Barnett, W. S., & Epstein, A. (1993). *Significant benefits: The High/Scope Perry Preschool study through age 27*. Ypsilanti, MI: High/Scope Press.

Schweinhart, L.J., & Weikart, D. P. (1980). *Young children grow up: The effects of the Perry Preschool program on youths through age 27*. Ypsilanti, MI: High/Scope Press.

Thompson, M.S. (1980). *Benefit-cost analysis for program evaluation*. Beverly Hills: Sage.

Triplett, J. E. (1992, April). Economic theory and BEA's alternative quantity and price indexes. *Survey of Current Business*, 72, 49-52.

U.S. Bureau of Economic Analysis. (1993). *Survey of current business*. Washington, DC: Department of Commerce, Bureau of Economic Analysis.

U.S. Bureau of the Census. (1983). *Lifetime earnings estimates for men and women in the United States: 1979*. (Current Population Reports, Series P-60, No. 139.) Washington, DC: USGPO.

Viscusi, K. (1986). The valuation of risks to life and health: Guidelines for policy analysis. In J. D. Bentkover, V. T. Covello, & J. Mumpower (Eds.), *Benefits assessment: The state of the art* (pp. 193-210). Boston: D. Reidel Publishing.

Young, A. H. (1992, April). Alternative measures of change in real output and prices. *Survey of Current Business*, 72, 32-46.

Chapter 13

IMPROVING THE USE OF EVALUATIONS: WHOSE JOB IS IT ANYWAY?

Carol H. Weiss

Evaluations wear blue jeans and work shirts and set out into the world to accomplish practical ends. Where research dresses up in a suit and tie to add to the world's knowledge, evaluation goes out to analyze the operations and outcomes of social programs and policies. Its aim is intensely practical: to improve program practice in the workaday world.

Yet experience has demonstrated that actual use of the results that evaluation produces is uneven. Some evaluators have reported great success in turning around weak programs, but other evaluators have recorded heartfelt laments that their work was ignored. How bleak or sunny the story is depends in part on what we mean by "use."

Some writers discuss "use" of evaluation results as the implementation of program changes based on shortcomings revealed by the evaluation. That is a rigorous definition indeed. For such use to occur, a long series of intermediary developments have to occur. Let's see what has to happen and where along the path "use" can break down.

Advances in Educational Productivity, Volume 7, pages 263-276.
Copyright © 1998 by JAI Press Inc.
All rights of reproduction in any form reserved.
ISBN: 0-7623-0253-4

Practitioners may not hear about the results of an evaluation at all. They may hear about the results but not understand them. They may understand them but not believe them. They may believe them but not know what to do about them. They may know what could be done but believe that *they* cannot do anything about them, because they lack either the authority, the resources, or the skills to make changes. Or they may not take action because they are satisfied with the way things are going now. In a good scenario they may start to take action, but then they run into roadblocks and grind to a halt. Only if many elements fall into line will they understand the implications of the findings for action, have the necessary resources for action, and successfully take action.

In this chapter, I will consider that evaluation results are "used" if practitioners get to the stage of knowing about and believing evaluation results and considering what to do. Use of evaluation is relatively simple when the changes that are required are simple, cheap, within teachers' existing repertoire, consonant with prevalent organizational practices and policies, and do not run afoul of political or ideological commitments. But when the implied changes are more controversial or far-reaching, even my relatively generous definition of "use" can become problematic.

USE FOR PRACTICE

This chapter is about the use of evaluation results for practice. Policy uses, while kissing cousins, have a different profile. In the case of policy, decisions tend to be made by a confined series of actors. The numbers of actors can run into the hundreds, of course, as administrative agency staff, legislators, district officials, school board members, advocacy organizations, unions, and/or community groups become involved. But around any particular policy decision, potential actors can usually be identified. Second, a policy is made over an identifiable period of time. It is not always made in a specific place on a specific date, because policies often take shape gradually as issues are debated and discussed, options considered and discarded, paths opened and other paths foreclosed. But at some point, a policy is affirmed. Third, a policy is expected to have long-term consequences. It is an official statement of direction that is expected to guide the actions of many people over an extended period of time. Fourth, although policies can occasionally be disregarded, especially in the absence of enforcement mechanisms or incentives and penalties, they are not lightly reversed. Official action usually has to be taken to overturn prior policies or to set new ones.

Practice, on the other hand, is cut to a different pattern. Educational practice is carried out by millions of teachers, administrators, and collateral personnel. It is ongoing, without definite benchmarks of time or achievement, not readily known by outsiders, rarely monitored outside the confines of a single school—and often not even there. Evaluation information can be "used" partially, in fragments, intermittently, inappropriately, not at all, or conscientiously, and then it can as easily be abandoned. The attempt to get evaluation information used by teachers has to contend with the amorphousness of the process, as well as with the scope of the task and the difficulty of discovering the extent to which it is in fact successful. This chapter will stress the use of evaluation by practitioners, although some of what I say has relevance for actors in policymaking as well.

WHAT SHOULD BE USED?

Many people who write about the use of evaluation findings construe the issue as a matter of how to *increase* use. These people, often evaluators themselves, make the assumption that evaluation has something useful and important to say and that if practitioners fail to heed its lessons, they are in that sense failing in their jobs—or at least failing to do the best that they can do with the resources at hand. Teachers do not necessarily make the same assumptions. They often contend that evaluation evidence does not apply to them and their classroom, that it is irrelevant or incomplete, that it misinterprets what is going on, that it is done by outsiders who do not understand the realities of the "classroom crucible" (Pauly, 1991), or even that evaluation data are outright wrong. Although I hate to acknowledge it after all the fine prescriptions in the earlier chapters in this book, the teachers have a point. Some evaluation is partial, irrelevant, outdated, inappropriate, or relevant only to a restricted subset of cases Attention to the findings would not necessarily improve teaching or administrative practice.

Therefore, this chapter is about *improving* the use of evaluation. I want to deal with the use of the best, most valid, and most relevant data available. Such use would not always be use of the most recent study of a particular program. Sometimes it would be attention to an integrated review or meta-analysis of prior evaluations. Sometimes it would involve teachers' collaboration in re-analyzing study findings. Sometimes it would actually mean ignoring findings of a fatally flawed study or a study that failed to take account of important features of the local situation. This chapter considers the issue of bringing the most appropriate evaluation results to attention in the context of educational practice.

Furthermore, evaluation does not provide all the answers to questions of practice. It does not deal with the myriad issues that teachers learn to deal with in the course of their education and experience. Evaluation does not often consider issues of the organizational environment, the community context, the matter of cost, demands from parents, support from other institutions, and so on. Practice has to take into account many conditions of which evaluation is only distantly aware.

None of this is meant to overlook the fact that much avoidance of evaluation findings springs from less high-minded sources. Teachers and administrators, like all other personnel whose programs are evaluated, can ignore important evidence because of defensiveness, apathy, or other self-serving motives. Using new information takes time, effort, skill, and attention, and practitioners are often more comfortable doing what they have always done in the ways they are accustomed to doing them. Let's recognize their reasonable skepticism about the merit and worth of some evaluative work without justifying their inattention to relevant and valid data.

THREE WELL-RECOGNIZED MEANS FOR IMPROVING THE USE OF EVALUATIONS

If evaluation is going to contribute to educational reform, evaluators have to do first-quality studies and communicate the results to people who can use them. In addition, it often helps to place the results in the context of evaluations of similar programs. An integrated review (or meta-analysis) of similar evaluations gives the audience a sense of how typical or unusual the current results are, and it can help to isolate the specific features of the program that were associated with greater or lesser effectiveness. Further, this kind of synthesis provides important information to educational institutions that were not the subject of the current evaluation but wish to learn whether they should implement programs of the same sort or improve they way they run their own programs.

High Quality Evaluation

The first and most important prerequisite for satisfactory use of evaluation results is a good study.[1] Acting on the basis of faulty evaluation would hardly be a contribution to education. Only if the study produces valid information is it likely to lead to program improvement.

Evaluation is a demanding enterprise, both a science and an art, and doing a first-class evaluation is a difficult feat even under good conditions. It requires capable and experienced evaluators, who have good knowledge of the programming field, sufficient funding, a long enough time period for study, and cooperative program personnel. It requires conditions that allow for repeated data collection, up-to-date and accurate program record keeping, and randomized assignment to control groups where necessary. Where these conditions are in short supply, or where political pressures demand premature release of results, suppression of unfavorable evidence, use of inappropriate comparisons, or other no-no's, the evaluation has to struggle even harder to produce cogent conclusions. To improve the use of evaluation, a good evaluation is the first principle.

Excellent Communication

A second basic strategy for improving the use of evaluation results is making them well and widely known. Many evaluators have devoted their energies to writing understandable reports, making them short and clear, illustrating them with attractive charts and illustrations, relegating statistical models and complex tables to an appendix, and producing executive summaries that can be absorbed while the reader stands on one foot. They have disseminated them widely, often with extra funding from granting agencies.

Further, they have recognized that the written word is a feeble device for reaching practitioners, and they have supplemented reports with oral presentations. They have talked with practitioners at the program site. They have offered training sessions and workshops. They have talked with administrators and district officials, school boards and parents associations. Nor have they limited their discussions to the period after the study was over. Some evaluators have given interim reports, both written and oral, while the study was in progress, in order to acquaint practitioners with emerging findings and avoid surprises at the end.

Syntheses of Evaluations of Similar Programs

Another general aid to evaluation use is meta-analysis. Whereas the single study may be highly relevant to the program that was evaluated, it is not necessarily the best guide for other schools or school systems. It is the captive of a particular set of events—the program, participants, staff, school, historical period, and community from which the results derived. Its results may not generalize well to other times and places. Fortunately, a remedy for this particularism is at hand. Meta-analysis synthesizes evidence from dozens, even hundreds, of studies that have been conducted of the same kind of program over a range of conditions and time periods. Each study becomes a data point in the meta-analysis, and the analysis reveals the effects of all the different implementations of the same kind of program. The oddball results of any particular study are mixed in with the results of other studies, and a more generalizable account is constructed.

Furthermore, by looking across studies, the meta-analyst has the opportunity to analyze the extent to which any specific program feature is associated with better outcomes. Thus, for example, the meta-analyst can discover whether programs of math instruction were more successful when they kept the same teacher with the class for multiple years, or organized the classroom into cooperative learning units, or offered after-school assistance (Cook et al., 1992). (See Becker's chapter in this volume for more information on meta-analysis.)

The key advantage of meta-analysis is the summation of available data from a wide variety of implementations of the same program. The idiosyncrasies of any individual program are set in the context of a representative accounting of how such programs fare under a range of circumstances.

THREE FURTHER STRATEGIES FOR
IMPROVING EVALUATION USE

Historically, three additional strategies for improving the use of evaluation results have dominated the literature. They arose in this chronological order:

- Intensive dissemination
- Evaluator-practitioner cooperation
- School self-evaluation ("empowerment" evaluation)

Intensive Dissemination

Throughout the history of evaluation, the major strategy used to improve (or increase) the use of results has been intensive dissemination. Dissemination strategies take multiple forms. In its most common form, it means following the prescriptions for excellent communication, and distributing the reports to all the potential users of the information. Great emphasis is placed on moving from abstruse and obtuse social science writing to understandable English narrative. Certainly, ensuring that program people can understand what the results *are* is an important precondition of use. However, by itself it is not sufficient to ensure utilization.

In the interests of utilization, evaluators have paid attention to the timing of reports. Releasing reports after important decision have been taken or commitments made is a barrier to anyone's use of results. Timeliness is a useful feature of a report dedicated to utility. But surprisingly, a late report does not foreclose all opportunities to have results used. If an issue is important and/or widespread, it comes up for consideration again and again. A report too late for influence in one setting can be used in other settings where similar programs are being run or considered for adoption, and it can even be used in the original setting during the next round of reform. Nevertheless, timely reporting is a virtue—and an aid to utilization.

Special strategies have been promoted and occasionally adopted. For example, Snare (1995/1996) has suggested that evaluators distinguish between potential users who are "pragmatists" and those who are "crusaders," and tailor their work to the predispositions of the different audiences. Shavelson (1988) has advocated the study of the "action mind frames" of practitioners, so that evaluators and researchers can design and conduct their studies and translate their findings into the frames of the designated audiences.

Writers have also put a lot of stock in easing access to evaluation information. They have advocated teaching practitioners to locate evaluation literature, helping them understand basic statistics, and setting up databases that are user friendly and readily available in libraries and educational institutions. A National Academy of Sciences panel, for example, urged the establishment of an electronic network that provides access to "research and exemplary practice information" (Atkinson & Jackson, 1992, p. 7). The assumption here is that practitioners would be interested in evaluation findings if only they knew where to look and could locate sources readily.

During the 1970s, the federal government sponsored special efforts to improve the use of evaluation and research in education. An innovative strategy was to employ "knowledge utilization agents," that is, middlemen between the practice and research communities who could locate the type of information that educational practitioners wanted and provide it to them. (Sieber et al., 1972, Louis et al. 1981). The agents were trained to seek practitioners' requests, scour the relevant databases for the desired information, and disseminate it to the targeted audience. Although the programs had some success in increasing the flow of research information, they were expensive and they did not always fit well into the organizational structures of state departments of education. When federal funding ended, the programs dwindled.

Evaluators over the years have developed a repertoire of other dissemination methods. They have expanded their reporting to include widespread mailings of written materials and presentations to large audiences of practitioners. Particularly when the study was national in scope, evaluators have spoken at national association meetings, such as the National Association of Secondary School Principals, and written articles for practitioner journals. On occasion, the mass media have picked up their findings, and a few evaluators have appeared on radio and television talk programs.

However intensive and well-calculated the efforts at dissemination have been, and some of them have been models of effective communication, they all contain the implicit assumption that the evaluator knows some truth that practitioners do not know and should know, and that she can tell them about it. A further assumption seems to be that new knowledge will lead to change in behavior. This cognitive assumption is widespread in many fields, namely that new learning will change people's understanding, which will lead to changes in their attitudes and commitments, and from there to alterations in their practice. It is an assumption that has not held up under many conditions, as evaluators of such behavior-change programs as pregnancy prevention and substance abuse prevention have frequently reported. Only occasionally does an increase in knowledge seem to be enough to stimulate changes in long-accustomed modes of behavior. As evaluators of behavior-change programs have themselves often found out, social supports are often required to reinforce and sustain behavioral changes and new resources may be required as well. It is also important that the target of the program have strong motivation and commitment to doing better. If we think of the use of evaluation results as a behavior-change program, which it is, then evaluators can profit from these lessons in their own domain.

Implicit in the intensive dissemination strategy, too, is a disregard for the special knowledge that practitioners can bring to the table. The next strategy explicitly takes the "wisdom of practice" into account.

Evaluator-Practitioner Cooperation

A main premise underlying closer cooperation with practitioners is that if teachers and principals have more of a say in an evaluation, they will come to feel a sense of ownership in it. They will regard it as at least partly theirs, they will take its findings seriously, and therefore when the study is over, they will make efforts to apply the results to their practice (Weiss, 1988a, 1988b).

Sometimes known as stakeholder evaluation (see the chapter by Alkin and colleagues in this volume), evaluations in the past 20 years have often attempted to include teachers and principals in the conduct of the evaluation. Sometimes the effort is concentrated at the beginning of the evaluation, when practitioners are asked about what issues and questions the evaluation should address. Other stakeholders, such as district officials, parents, and the outside program funder, may be included in the meetings and conversations. The evaluator then tries to direct the study to the matters that are most important to local people. But such canvass of the priorities of local stakeholders is not an original development of the past two decades; it has been part of the repertoire of applied researchers for at least 50 years (Lazarsfeld & Reitz, 1975; Coleman, 1972). What is new is the deeper involvement of stakeholders in other phases of the evaluation process.

Most efforts at improving utilization through collaboration occur at the end of the study when the data are ready. The evaluator presents the data, and asks teachers, principals, and other stakeholders to help interpret the meaning and implications of the data. Perhaps the evaluator presents her own provisional interpretation of the data, but she is open to alternative ideas and explanations of what the data mean. Such a procedure gives the evaluator an opportunity to learn as well as teach, to engage in discussion more than oration, and to enrich her understanding of the program in its setting.

Discourse of this kind can go on over extended periods of time. Huberman (1989) has written about "sustained interactivity," discussions between evaluator and teachers maintained regularly over a year or more. Together they confront issues about how to use the evaluative information in the concrete instances of classroom practice. The evaluator is one participant in the search for practical means to implement the information that the evaluation revealed, and often she has to reconsider the interpretation of the information in light of the problems that arise in the course of using the information in practice. In the process, the evaluator learns about practice in ways that are likely to enrich her subsequent evaluations, just as the practitioners learn much about evaluation.

Some supporters of evaluator-practitioner cooperation advocate deeper involvement for practitioners. They believe that teachers and other school people should take an active part in all the aspects of the study—not only helping to set the central issues for evaluation at the beginning and helping to interpret the meaning of the analyzed data at the end, but also framing specific questions and measures, collecting data, and analyzing the data. The potential users of the evaluation are full partners in its conduct. Therefore, they should feel a deep obligation to put its findings into practice to improve education.

The involvement of stakeholders in evaluation has other purposes besides increasing their use of evaluation. It is also meant to be more democratic than usual research relationships, that is to reduce the power differential between the studier and the studied. It is intended to give staff a greater sense of confidence in their own efficacy. It is sometimes expected that by giving them exposure to the methods and procedures of evaluation, teachers will become more favorably inclined toward evaluation and accountability. There are also those who hope that cooperation in evaluation will help school people to maintain better records and undertake some modest evaluative activities on their own.

Considered only on its value *as a means to* increase the use of evaluation, evaluator-practitioner cooperation rests primarily on the premise that participation will encourage school people to buy into the results, and because the results are in some sense "theirs," they will try to put them into use. Experience suggests that there is some merit in this expectation. Evaluations in which stakeholders have played an active part often do show greater efforts at implementing the results. However, outside conditions can interfere with this result. Staff who were involved in the evaluation not infrequently leave the school or change jobs before the results are ready, and their replacements have no commitment to the study. On occasion, disagreements and ill feeling between the teachers who took part in the study and those who did not can reduce the application of results. If implementation of changes in practice require more money, or structural changes in the organization of the school, or district office agreement, or union assent, or parental support, the buy-in of teachers may not be sufficient to carry the day. Of course, stakeholder evaluation can seek to include all these interests in the conduct of the evaluation. In fact, it is usually advisable to have principals, district office officials, union representatives, and perhaps parents participate in the planning and review of the evaluation, so that the concerns and knowledge of the whole panoply of actors are represented. But the representation of multiple groups makes reaching consensus about the study more difficult and complicates the effort to draw prescriptions from the data.

School Self-Evaluation

Some evaluators believe that stakeholder evaluation does not give school people enough power over the evaluations that may affect their lives. They recommend that school staff undertake the evaluations themselves. Staff can call in an evaluator to give them an orientation to evaluative practice and sage advice about how to proceed, but control of the study should be in their hands. If they need further help, they can turn to the evaluator, but she is a coach on the sidelines rather than a player. She waits until her help is requested before taking part. In this view, it is more important to "empower" the staff than to mount a study that has all the niceties that professional evaluators treasure.

This kind of "empowerment evaluation" (Fetterman et al., 1996) does not have utilization as its primary rationale. It rests on a political argument about the relative power of evaluators and practitioners and the need to right the balance. It also takes for granted that teachers have specialized knowledge about teaching and students that are important components in any realistic evaluation. Nevertheless, a prevailing expectation is that teachers' own evaluation of a school program will definitely be acted upon. Nobody has to peel away the layers of suspicion that teachers have about outside evaluators, and nobody has to sell them the importance of the information revealed. This is their own study, and they will be committed to make it the basis for change.

Empowerment evaluation has much in common with the movement that is called "teacher research." A number of educational researchers—and teachers—are advocating that teachers undertake their own research in their own classrooms, rather than serve as passive subjects of other people's research. In this view, teachers can create knowledge that is relevant to their own questions, with their own students, in their own organizational contexts. Instead of being the subjects (sometimes reluctant) or the recipients (often skeptical) of research developed by university researchers, they themselves construct knowledge. Such research will have immediate applicability—and application. The teacher research movement is a research analog to empowerment evaluation.

Empowerment evaluation is a relatively recent phenomenon, or at least a newly named phenomenon, but its predecessor—the organizational self-study—goes back to the early days of applied social research. Not a great deal of experience has accumulated about its recent incarnation. It no doubt has the virtues of evaluations developed through evaluator-practitioner cooperation, perhaps to an even greater level. School people do in fact "own" the evaluation. It is their study, responsive to their questions. It fits the local situation. But it probably has additional disadvantages. Because laypeople are doing the evaluation, the methods of study are not likely to be as rigorous as evaluations done by professional evaluators. Moreover, the topics that school people choose to study are likely to be relatively innocuous. Teachers and principals, without inputs from an evaluator or other stakeholders, are apt to want to evaluate a program in terms that they can do something about within the existing patterns of authority. In many ways, of course, that is an advantage for utilization; they study matters that they have the power to change. But for the improvement of education, it can be a limitation. They are likely to confine the study's attention to relatively small-scale practices and not question the allocation of resources, the division of authority, or the policy mandates under which they work. They are not likely to question matters that fall beyond their purview, and thus the evaluation may produce limited information.

Advocates of the approach cite successes, and in certain circumstances, it can probably be a useful way to go. But I would hope to see more emphasis on rigor in the discussions of self-evaluation, whether practitioners undertake quantitative or qualitative investigation. Qualitative inquiry demands as much professionalism in data collection and analysis as any quantitative study, and it is usually harder to do well. It would be a shame to encourage practitioners to do slipshod studies, especially given all the other demands on their time. Even if such evaluations were used, they would probably not do much to improve the state of education. On the other hand, practitioner engagement in evaluation can have advantages. The most important contributions are likely to be (1) the first-hand knowledge that school people gain from empirical inquiry into conditions and events and (2) their exposure to the process of inquiry, to the questioning state of mind, that proceeds on the basis of evidence rather than faith. The multiplication of reflective practitioners (Schon, 1983) is not an inconsiderable benefit.

I have discussed three strategies for improving the utilization of evaluation in this section—intensive dissemination, evaluator-practitioner cooperation, and staff self-evaluation. Each has merit under appropriate conditions, but none by itself is enough to guarantee that evaluation results are seriously taken into account for serious change. What they can accomplish is help move toward that end. For evaluation results to be influential in school practice, both evaluators and practitioners have to bring something to the table, too, their own covered dish to the collective party. Evaluators have to bring not only their research skills but their responsiveness to practitioners' questions and perspectives and their communication ability. Practitioners have to provide their first-hand awareness of the issues involved and—above all—the will and determination to change what is wrong.

A FOURTH STRATEGY: INCREASES ON THE DEMAND SIDE

All of the previous techniques for improving the use of evaluation rely on reforming the modes of evaluation or its presentation. The work is all on the supply side of the evaluation-practice equation. It is assumed that evaluators need to change what they do and how they do it. But what about the demand side? Why isn't there a raging demand on the part of school people for information that will help them do their work more effectively? Why don't they demand full and frequent information about the effectiveness of their processes and the outcomes of their work so that they can improve the quality of what they do? An alert elementary school teacher with a single class can probably see the consequences of her work quite directly, but the same cannot be said of teachers in higher levels of schooling who deal with many different classes daily, supervisors of services such as vocational, special education, or bilingual, chairs of departments, or administrators of schools, colleges, and universities. If those of us in such positions were motivated to do the best possible job, we would not only welcome but painstakingly search for the most valid and relevant information available.

The current popularity of the concept "the learning organization" derives from the realization that in today's changing world, all organizations must continuously learn. As Senge (1990), Argyris and Schon (1996) and others have written, the world is too complex and too fast-moving for any organization to rest on the laurels of the past. (And not all organizations have past laurels to lean on.) It must constantly draw in new knowledge and ideas in order to keep up with change.

If any organizations would seem to have an advantage in becoming "learning organizations," it would be schools. Learning is what schools are all about. However, most schools think of themselves as places where young people learn but adults make do with what they already know. For schools to improve the education of the young people in their charge, adults have to keep learning new ideas, skills, and information that will help them lead to the ongoing process of school reform.

The many failures of evaluation utilization can not all be laid at the door of the evaluator and her inadequacies as researcher, communicator, or collaborator. The potential audiences for evaluation should be on their feet searching for good information and demanding the best data possible. When evaluations are ignored, the failure may be less a matter of an evaluator's faulty communication than of educators' inadequate motivation for improving the quality of their daily work.

By extension, some of the problem is apt to be inadequate management. Managers of schools and other educational organizations have a responsibility to motivate their members to achieve at the highest possible levels. They should be providing the resources and the opportunities to help them do so and engaging in collective efforts to harness everyone's talents to the task. Usually, good evaluative information is one of the most useful resources. Therefore, management guided by a search for excellence would be an eager customer of relevant evaluation.

CONCLUSION

Evaluators have often worked hard to increase the appropriateness, relevance, and quality of the studies they do and to disseminate the results widely to people who could profit from their lessons. They can do more. Some need to learn what the salient issues are *before* they begin their study, canvassing a range of viewpoints and not settling for the mandate of the study's sponsor. Some need to find ways to speak to practitioners on a frequent basis and to listen as well as tell. Some need to write more engagingly, exchanging the multisyllabic vocabulary of the research institute for the clear prose of the world of work. Many would find it useful to collaborate with expected users of evaluation findings in the planning of the study and in the interpretation of the data in light of the concrete issues facing practitioners. On occasion, they can collaborate even in the implementation of the evaluation, engaging practitioners in the ongoing data collection and analysis.

However, there are limits to what can be accomplished by trying to "sell" evaluation findings to a reluctant audience. Especially when the findings point out shortcomings in what practitioners have been doing, many of them all too humanly turn a deaf ear—or react with defensiveness. More wholehearted acceptance of the range of evaluation results—and their implications for practice—can be achieved only when practitioners realize that it is in students' best interests, and their own, to pay attention. When all of us in education are motivated to provide the best possible education for all students, we will look to many sources and resources for guidance, and good evaluation will be high on the charts.

NOTE

1. In this connection, let me note that evaluations can provide data about both program process and program outcomes. Both kinds of information can contribute to educational improvement.

REFERENCES

Argyris, C. & Schon, D. A. (1996). *Organizational learning II: Theory, method, and practice*. Reading MA: Addison-Wesley.

Atkinson, R. C., & Jackson, G.B. (Eds.). (1992). *Research and education reform: roles for the office of educational research and improvement*. Washington DC: National Academy Press.

Fetterman, D. M., Kaftarian, S. J., & Wandersman, A. (Eds.). (1996). *Empowerment evaluation: Knowledge and tools for self-assessment and accountability*. Thousand Oaks CA: Sage.

Huberman, M. (1989). Predicting conceptual effects in research utilization: Looking with both eyes. *Knowledge in Society: The International Journal of Knowledge Transfer*, 2, (3), 5-24.

Louis, K.S. (1981). External agents and knowledge utilization: Dimensions for analysis and action. In R. Lehming & M. Kane (Eds.), *Improving schools: Using what we know*. Beverly Hills CA: Sage.

Morrissey, E. Wandersman, A., Seybolt, D., Nation, M., Crusto, C., & Davino, K. (1997). Toward a framework for bridging the gap between science and practice in prevention: A focus on evaluator and practitioner perspectives. *Evaluation and Program Planning, 20*, (3), 367-377.

Pauly, E. (1991). *The classroom crucible: What really works, what doesn't, and why*. New York: Basic Books.

Reimers, F. & McGinn, N. (1997). *Informed dialogue: Using research to shape education policy around the world*. Westport CT: Praeger.

Schon, D.A. (1983). *The reflective practitioner*. New York: Basic Books.

Senge, P. M. (1990). *The fifth discipline: The art and practice of the learning organization*. New York: Doubleday.

Shavelson, R. J. (1988). Contributions of educational research to policy and practice: Constructing, challenging, changing cognition. *Educational Researcher, 17*, 4-11, 22.

Sieber, S. D., Louis, K. S., & Metzger, M. (1972). *The use of educational knowledge: Evaluation of the pilot state dissemination program*, Vols. I & II. New York: Columbia University Bureau of Applied Social Research.

Snare, C. E. (1995/1996). Windows of opportunity: When and how can the policy analyst influence the policymaker during the policy process. *Policy Studies Review, 14*, (3/4),407-430.

Weiss, C.H. (1988). "Evaluation for decisions: Is anybody there? Does anybody care?" *Evaluation Practice, 9*, (1), 5-19.

Weiss, C. H. (1988b). If program decisions hinged only on information. *Evaluation Practice, 9*, (3), 87-92.

Chapter 14

TRENDS IN EVALUATION RESEARCH

Herbert J. Walberg and Arthur J. Reynolds

Though evaluation research is still a young field, its evolution has been remarkable. Its origins may be traced to a skepticism engendered by academic and program theories that were expected to but did not work in practice. In the 1960s, the powerful influence of hard applied sciences, particularly agricultural and medical research, led applied psychologists to an emphasis on experiments to detect possible effects of educational and other innovations on learning and other outcomes. However, the requirement of random assignment to treatment and control groups offered difficulties. It proved hard to get people to do what a coin flip dictated, especially among large, varied, samples of people in ordinary settings who think and act for themselves. The circumstances of experiments sometimes appeared contrived, moreover, which cast doubt on whether any program effects would survive the vicissitudes of reality.

EVOLUTION OF EVALUATION RESEARCH

For these reasons, Donald Campbell, Julian Stanley, and others designed quasi-experiments, which allowed comparisons of non-randomly assigned program and control groups to evaluate program effects. Though not foolproof, quasi-experiments were far easier to conduct. They were less intrusive and could

Advances in Educational Productivity, Volume 7, pages 277-283.
Copyright © 1998 by JAI Press Inc.
All rights of reproduction in any form reserved.
ISBN: 0-7623-0253-4

be more widely implemented in many, varied conditions, which allowed wider generalizations. Quasi-experiments also allowed greater design flexibility and, as William Corrin and Thomas Cook show in their chapter, greater attention to the elements of designs can promote better evaluation research.

As described by Marvin Alkin, Carolyn Hofstetter, and Xiaoxia Ai in this book, people who would be affected by program decisions could more easily be engaged in the design and conduct of evaluations. As reflected in Carol Weiss's chapter, practitioners and policy makers that are engaged in this way may be more likely to use evaluation results. They may also come to see evaluation information as a valuable educational resource.

In the 1970s, quasi-experimental comparisons attracted economists, psychologists, sociologists, and statisticians whose practice it was to draw causal inferences from non-experimental data. They estimated program effects on groups that had been naturally rather than randomly assigned to various treatments and non-treatments in a variety of circumstances. As exemplified in Judy Temple's chapter, findings from such evaluations allow a broader basis of inference, that is, about how programs work with different groups under various conditions.

At the same time, surveys of large national samples of educational achievement, attitudes, and their possible causes became available for evaluation and policy analysis. Following precedents in economics and sociology, secondary analysts studied the apparent influence on individuals of naturally occurring educational and social conditions. Rather than accepting membership in treatment and control groups alone as possible causal determinants, they measured degree of exposure to specific aspects of educational programs and simultaneously occurring conditions. In their chapter, Robert Boruch and George Terhanian show how estimates of effects from controlled experiments can be linked to those from national surveys samples through cross-design synthesis as well as propensity score analysis.

As David Rindskopf's chapter illustrates, new analytic techniques allow more valid causal inferences from the usual evaluation data that are flawed in one way or another, particularly in being non-experimental. Just as epidemiology and public health studies complement experimental medicine, analytic approaches such as selection modeling, propensity score analysis, and growth curve modeling strengthen traditional program evaluation and may have significant implications for legislative policy.

Rich data sets allowed multivariate analysts to estimate the multiple influences of several factors on outcomes. Inspired by economics, for example, educational productivity research examined the joint influences of home environments, peer groups, mass media, schools, instruction, and individual differences among students on achievement and other outcomes. In such analyses, secondary analysts traded some causal certainty for wider derivation of results from multiple and varied "real world" settings.

In addition, secondary analysts gained another advantage: Program evaluators had traditionally employed covariance analysis to remove the effects of differences among students and conditions affecting outcomes such as socioeconomic status or parental support of learning so that the "purer" effects of programs could be measured. Instead of submerging such variables, regression analysts include such variables in their equations not only to control for but to study multiple causes. In this way, they gained more information about the range of possible influences. They found, for example, that conditions such as parental support were candidates for new programs and additional research.

By the 1980s, many evaluations, policy analyses, and related studies had accumulated. Dozens of experimental and quasi-experimental studies on some programs and related research topics demanded meta-analyses to synthesize their collective findings. Starting in education and psychology, meta-analyses multiplied and grew in influence on programs and policy as shown by articles in practitioner journals and in such new movements as "evidence-based education" and "evidence-based medicine." Meta-analysts often brought together the results of experimental and quasi-experimental comparisons of groups. In this book, Betsey Jane Becker describes techniques for synthesizing multivariate (multiple-influence) secondary analyses of educational productivity. Though such productivity studies may incorporate program comparison, they usually assess the effects of degrees of exposure to several educational methods and conditions.

At the same time, doubts remained or even increased about the effects of expensive programs. As Steven Barnett asks, do their program effects or benefits outweigh their costs? In some instances, apparently not: Control groups have sometimes performed as well as or better than those in new, complicated, and expensive programs. One reason, emphasized by Huey Chen, is defective program theory. As Ann McCoy and Arthur Reynolds document, moreover, prerequisite conditions for programs and the programs themselves may not have been in place. As illustrated by Jennifer Greene's and Linda Mabry's chapter, interpretive and case-study evaluations are informing program designers, administrators, and policy makers about the reasons for program success and failure.

These various methods evolved and were sometimes combined in particular evaluations. As they were, evaluators became more eclectic and cosmopolitan. Some appeared to blur usefully the distinction between program evaluation and policy analysis (as represented in the title of the American Educational Research Association's journal, *Educational Evaluation and Policy Analysis*). Programs might be considered different embodiments or particular manifestations of broad national policies carried out at the local level. Both evaluators and policy analysts, for example, might raise questions about the general effects of Chapter 1 for children from poor families as a national policy or about the comparative effects of specific Chapter 1 programs or program variations. Greater attention to strategies for disseminating evaluation findings and incorporating them in policy decision making also has occurred. In all of these ways, evaluation research broadened its scope.

CURRENT TRENDS

The previous chapters show the wisdom and utility of four emerging trends: (1) tailoring evaluation to program and policy questions, (2) cumulating and integrating information, (3) collaborative research, and (4) comprehensive efforts. These trends seem likely to continue and to contribute to the design of more productive educational and social programs. Each deserves a few words in conclusion.

Tailoring Evaluation to Policy Questions

Evaluation research may be viewed as a question-driven enterprise. In his chapter, Huey Chen distinguishes questions about program impact from program implementation. He argues that experimental and quasi-experimental methods appear more appropriate when chief question concerns the impact of programs; theory-driven evaluation is more appropriate for investigating how programs operate. Similarly, the stakeholder approaches discussed by Marvin Alkin, Carolyn Hofstetter, and Xiaoxia Ai incorporate key questions that target audiences want addressed, for example, regarding staff and consumer satisfaction. As Alkin and his colleagues point out, stake holding collaborators are more likely to accept and use findings.

In her chapter on qualitative evaluation, Jennifer Greene notes that answering the question "What is the overall effectiveness of this educational innovation?" usually necessitates experimental or quasi-experimental methods. On the other hand, such questions as "In what ways does this new curriculum promote intended problem solving skills among participating students?" suggests a qualitative approach. Identifying the most important evaluation questions helps direct evaluators to the most appropriate methods.

As Steven Barnett illustrates, economic analyses should also be influenced by the questions most salient to stakeholders. Evaluators have usually been concerned with the kinds of questions just raised: What works? How does it work? For whom does it work best? Practitioners and policy makers may be even more concerned with the question, "What are the program costs, possibly also considering estimated effects and monetary benefits of programs." With a fixed budget or little difference in program effects—both of which are common--cost questions loom even larger in their minds. Cost-effectiveness analysis and cost analysis are also valuable approaches for assessing such questions.

Cumulating and Integrating Information

Several chapters show that drawing information together from multiple sources can promote its utility. Boruch and Terhanian's cross-design synthesis incorporates both experimental and survey data to address evaluation questions of high significance such as the effects of ability grouping or adult literacy programs. William Corrin and Thomas Cook show how information obtained from comparison groups, matching, multiple pretests, repeated measures, nonequivalent dependent variables, and program replications can strengthen causal descriptions.

As illustrated in Judy Temple's chapter, a single secondary analysis may yield estimates of the simultaneous influences of several programs and conditions from national databases. Since such analysis is often based on large, random samples, it allows more secure sample-to-population inferences than the usual program evaluation samples, which consist of small numbers of program and control groups. Indeed, some of the most influential evaluations are secondary analyses such as the Coleman report on U.S. schools, Coleman's analyses of the effects of public versus private schools, and Cook and colleagues' study of the effects *Sesame Street*. Becker shows how multivariate meta-analysis can help synthesize findings from such multiple-cause surveys and research on educational productivity. Such secondary analysis and multivariate synthesis allow assessment of the consistency of program effects across different outcomes, samples, and settings.

As both Jennifer Greene and Linda Mabry illustrate in their chapters on qualitative evaluation and case studies, evaluation conclusions are qualitative judgements based on the quantitative and qualitative data in hand. Successful integration of qualitative and quantitative knowledge can lead to better understanding about the importance of educational programs and policy, and this certainly increases their salience in policy settings. As Ann McCoy and Arthur Reynolds indicate, the process of program implementation also benefits from the mixture of methods.

Collaborative Research

Any evaluator would find it difficult to master all the methods described in the foregoing chapters; each method requires specialized training and practical experience. For this reason, collaborative research is often the most practical approach to conducting comprehensive evaluations. Several evaluators, each with specialized expertise, can work closely together as a team. At longer distances, meta-analytic conclusions can inform the secondary analysts, and meta-analysts can draw together findings furnished by secondary analysts, as suggested in Becker's and in Temple's chapters.

Such explicit or implicit collaboration, represented in many of the chapters, is more common in evaluation practice today. Witness, for example, the triangulating of experimental research procedures with sample surveys and site-level case studies in such national evaluations such as the Comprehensive Child Development Program and the Head Start-Public School Transition Demonstration Program. As both Jennifer Greene and Linda Mabry show, qualitative and case study evidence may illuminate quantitative evaluations. As explained by Boruch and Terhanian, cross-design methods require integration of experimental and survey research data and procedures, which calls for collaboration of researchers experienced in experiments and in sample surveys. Likewise, as illustrated in Temple's chapter, secondary analysis of large national data sets often requires both substantive and statistical expertise; and the sheer volume of data calls for teamwork. Meta-analysts may face similar challenges, and as Becker discusses, meta-analysis has its own statistical literature. Evaluation researchers often require the advice or collaboration of statisticians and psychometrists.

Collaboration also may be valuable, if not critical, in theory-driven evaluations as discussed by Chen, since specification of alternative program theories benefits from knowledge about the program that evaluators rarely possess. They can learn much from program directors, staff, and administrators. As Alkin and colleagues describe them, moreover, stakeholder approaches are collaborative almost by definition. Similarly, participatory evaluation, discussed by Greene and in the Alkin and colleagues chapter, requires collaboration and has the advantage of promoting dissemination and use. By extension, educational systems may undertake self-evaluations ("empowerment" evaluations), which as discussed by Weiss, can also improve evaluation use but at the risk of addressing second-order questions.

Comprehensive Efforts

Theory-driven and stakeholder approaches, emphasis on program implementation, advances in quasi-experimental and experimental approaches, research synthesis, qualitative and case study approaches, secondary analysis, and strategies for improving use all have significantly contributed to the development of evaluation in the past decade. This book shows that evaluation concerns activities from program theory through dissemination and extending into policy analysis and utilization of findings. Each part of spectrum of evaluation helps to define the field more broadly than in past decades. Neglecting any one of these can lead to incomplete results or unrealized use. As McCoy and Reynolds indicate, evaluations of program implementation were often given low priority in the past and, as a result, many programs judged ineffective were just poorly implemented—a possibly correctable problem.

As Weiss details in her chapter, the impact of evaluation itself depends in part on how well it is communicated to policy makers and planners. While communication is necessary, it is not always sufficient to get evaluation results used. Stakeholder involvement in the evaluation can often help. In addition, utilization will be aided by (a) intensive dissemination to audiences in many locations, (b) ongoing cooperation with program staff during their implementation of evaluation results, (c) evaluators' willingness to listen and learn about the settings in which results will be applied, and (d) stimulation of the demand for evaluative information for purposes of program improvement and accountability. Of course, improving the accuracy of evaluation findings with the methods and perspectives described in this volume also may increase utilization.

CONCLUSION

Evaluation research should have educational and social benefits; it should lead to the design of more effective programs and policies. Better programs and policies should lead to more learning. Specifically, as suggested by the previous chapters, how can evaluators best contribute to educational productivity? First, evaluators can provide accountability. They can ask such questions: Are programs appropriately designed and implemented? Are program resources being used properly? Do programs efficiently achieve their intended effects. How do unsuccessful implementations fail? What are the program effects in relation to their costs? To the extent that answers to such questions help identify the best practices, educational productivity can be facilitated. As suggested in several of the chapters, we can expect that the most effective programs are based on research and evaluation findings and that their administrators continuously monitor and evaluate their efforts.

Second, evaluators can contribute to program development and policy improvements, but more than careful research is required. As Weiss suggests, they should give some of their energies and time to dissemination and helping policy makers and program administrators understand findings and draw their implications. Well-conducted evaluations do not inevitably lead to use, even with good communication; but when evaluations are used, the excellence of a study makes the consequences of use much more socially valuable.

Finally, evaluators can contribute more to educational productivity if they help plan and evaluate programs more comprehensively. Investigating implementation contexts and processes, identifying causal mechanisms of programs, determining who benefits most and why, determining the generalizability of intervention effects across different settings, and improving evaluation utilization—all of these can help increase educational productivity today and in the future. Persistent implementation of the methods and perspectives described in this volume should continue to advance the evaluation field to meet the challenges of tomorrow.

ABOUT THE CONTRIBUTORS

Arthur J. Reynolds is Associate Professor of Social Work and Child & Family Studies and Research Affiliate of the Institute for Research on Poverty at the University of Wisconsin-Madison.

Herbert J. Walberg is Research Professor of Education and Psychology at the University of Illinois at Chicago.

Xiaoxia Ai is a doctoral student in the Graduate School of Education and Information Studies at UCLA.

Marvin C. Alkin is Professor in the Graduate School of Education and Information Studies at UCLA.

W. Steven Barnett is Professor in the Department of Educational Theory, Policy, and Administration, Graduate School of Education at Rutgers University.

Betsy Jane Becker is Professor of Measurement and Quantative Methods, Department of Counseling, Educational Psychology, and Special Education in the College of Education at Michigan State University.

Robert F. Boruch is University Trustee Professor of Education and Professor of Statistics in the Wharton School at the University of Pennsylvania.

Huey-tsyh Chen is Professor of Sociology at the University of Akron.

Thomas D. Cook is Professor of Sociology, Psychology, Education and Social Policy, and Fellow of the Institute for Policy Research at Northwestern University.

William J. Corrin is a doctoral student in the Department of Sociology and Adjunct Lecturer in the Department of African-American Studies at Northwestern University.

Jennifer C. Greene is Associate Professor in the Department of Policy Analysis and Management, College of Human Ecology at Cornell University.

Carolyn H. Hofstetter is a doctoral student in the Graduate School of Education and Information Studies at UCLA.

Linda Mabry is Assistant Professor in the Department of Counseling and Educational Psychology at Indiana University.

Ann R. McCoy is Senior Research Associate in the Center for Academic and Reading Skills at the University of Texas-Houston Health Sciences Center.

David Rindskopf is Professor of Educational Psychology at the City University of New York Graduate Center.

Judy A. Temple is Associate Professor of Economics at Northern Illinois University and Visiting Scholar in the Institute for Research on Poverty at the University of Wisconsin-Madison.

George Terhanian is affiliated with the Graduate School of Education at the University of Pennsylvania.

Carol H. Weiss is Professor in the Graduate School of Education at Harvard University.

AUTHOR INDEX

SUBJECT INDEX

Advances in Educational Productivity

Edited by **Herbert J. Walberg,** *College of Education, University of Illinois at Chicago*

Volume 6, Optimizing Education Resources
1996, 281 pp. $78.50/£49.95
ISBN 0-7623-0082-5

Edited by **Bruce S. Cooper,** *Graduate School of Education, Fordham University* and **Sheree T. Speakman,** *K-12 Education Team, Coopers & Lybrand L.L.P*

CONTENTS: Foreword, *Herbert J. Walberg.* Introduction. Twenty-five Years Later: *Serrano* Goes to School, *Bruce S. Cooper and Sheree T. Speakman.* PART I. CASE STUDIES OF SCHOOL EQUITY AND EFFECTIVENESS: NEW JERSEY, HARTFORD, CT, AND BOSTON. The Quest for Education Equity in New Jersey: The Experience of Governor James J. Florio, *Warren P. Howe.* Cost-Effectiveness? Or Effectiveness at a Cost? EAI and the Hartford, Connecticut Public Schools, *Thomas H. Jones.* Freeing Resources for Improving Schools: A Case Study of Teacher Allocations in the Boston Public Schools, *Karen Hawley Miles.* PART II. THE BATTLE FOR FINANCIAL EQUITY: FEDERAL, STATE, LOCAL DISTRICT, AND SCHOOLSITE. Meeting the Challenge of Devolution: How Changing Demography and Fiscal Contexts Affect State Investments in Education, *Martin E. Orland and Carol E. Cohen.* The Assessment of Equal Educational Opportunity: Methodological Advances and Multiple State Analyses, *Deborah A. Verstegen.* Tracing School-Site Expenditures: Equity, Policy, and Legal Implications, *Sheree T. Speakman, David C. Bloomfield, Bruce S. Cooper, Jay F. May, Robert M. Sampieri, Hunt C. Holsomback, Larry Maloney, Judith Nappi, and Deborah Rosenfield.* PART III. IMPROVING SCHOOL PRODUCTIVITY FROM MONEY TO TEACHERS TO STUDENT LEARNING. Where's the Money Gone?, *Richard Rothstein.* Are Productivity and Equity Incompatible? The Case of Texas School Finance, *Lawrence O. Picus.* Educational Achievement and Teacher Qualifications: New Evidence from Microlevel Data, *Dominic J. Brewer and Dan D. Goldhaber.* The Demographics of School Funding Variations: An Attempt to Inform the Debate, *Linda Hertert.*

Also Available:
Volumes 1-5 (1990-1995) $78.50/£49.95 each